CO-BEH-862

Pesticides: Managing Risks and Optimizing Benefits

ACS SYMPOSIUM SERIES **734**

Pesticides: Managing Risks and Optimizing Benefits

Nancy N. Ragsdale, EDITOR
U.S.Department of Agriculture, Beltsville MD

James N. Seiber, EDITOR
U.S. Department of Agriculture, Albany CA

American Chemical Society, Washington, DC

Library of Congress Cataloging-in-Publication Data

Pesticides : managing risks and optimizing benefits / Nancy N. Ragsdale, editor. James N. Seiber, editor.

p. cm—(ACS symposium series : 734)

Includes bibliographical references.

ISBN 0–8412–3616–X

1. Pesticides—United States Congresses. 2. Pesticides—Environmental aspects—United States Congresses.

I. Ragsdale, Nancy N., 1938– . Seiber, James N., 1940– . III. Series,

S8950.2.A1P485 1999
363.17'92—dc21

99–14981
CIP

The paper used in this publication meets the minimum requirements of American National Standard for Information Sciences—Permanence of Paper for Printer Library Materials, ANSI Z39.48-94 1984.

Distributed by Oxford University Press

PRINTED IN THE UNITED STATES OF AMERICA

Advisory Board

ACS Symposium Series

Foreword

THE ACS SYMPOSIUM SERIES was first published in 1974 to provide a mechanism for publishing symposia quickly in book form. The purpose of the series is to publish timely, comprehensive books developed from ACS sponsored symposia based on current scientific research. Occasionally, books are developed from symposia sponsored by other organizations when the topic is of keen interest to the chemistry audience.

Before agreeing to publish a book, the proposed table of contents is reviewed for appropriate and comprehensive coverage and for interest to the audience. Some papers may be excluded in order to better focus the book; others may be added to provide comprehensiveness. When appropriate, overview or introductory chapters are added. Drafts of chapters are peer-reviewed prior to final acceptance or rejection, and manuscripts are prepared in camera-ready format.

As a rule, only original research papers and original review papers are included in the volumes. Verbatim reproductions of previously published papers are not accepted.

ACS BOOKS DEPARTMENT

Contents

BIOLOGICAL AND ECONOMIC BENEFITS

GLOBAL CONSIDERATIONS

INDEXES

Preface

Bring up the topic of pesticides and all too frequently the ensuing discussion will be filled with emotional issues, often extending well beyond the realm of science and heavily flavored with the stimulus of political gain. These adversarial arguments have been detrimental to an objective evaluation of the risks and benefits associated with pesticide use and the role pesticides play in modern society. The reduction of risks associated with pesticide use has seen significant advances over the past 30 years. To continue this trend, there should be a constant, scientifically-based evaluation of both risks and benefits, determining how risks can be reduced and how significantly pesticides support strong agricultural production systems.

The goal of the symposium that provided the basis for this book was to provide a forum in which the positive steps that have been, or could be, taken to reduce risks could be presented side by side with the benefits of pesticides. After planning for the symposium was underway, Congress unanimously passed the Food Quality Protection Act (FQPA), imposing changes in data required and how risks would be determined, not only for initial pesticide registrations, but also for maintaining currently registered uses. In view of this new legislation, a section on the FQPA was added to the symposium. Although the FQPA altered the role of pesticide use benefits in the regulatory process, this topic remained because efficient agricultural production systems, required by an ever-increasing population, mandate a consideration of use benefits from a practical perspective.

This book examines various aspects of pesticide risks and benefits, first giving a general overview of the topic. This is followed by a section devoted to the FQPA, because that will be the driving force for U.S. use of pesticides. The FQPA section outlines the statute, covers the major factors that will influence future risk determinations, and examines impacts at the state level. The next two sections look at factors involved in reducing or managing risks and at various considerations associated with pesticide benefits analyses. Ultimately a section is devoted to issues associated with pesticides in the arena of global trade.

The broad array of topics makes this book valuable to a wide audience, ranging from scientists to policy makers. Presentation of risk and benefit factors together make this publication somewhat unique. Emphasis is placed on the importance of taking both into consideration, whether addressing future research or pesticide policy. The editors express their appreciation to all the authors as well as the reviewers that generously gave their time and thoughts to making this a successful publication.

NANCY N. RAGSDALE
Agricultural Research Station
U.S. Department of Agriculture
Building 005, Room 331, BARC-W
Beltsville, MD 20705

JAMES N. SEIBER
Western Regional Regional Research Center
Agricultural Research Station
U.S. Department of Agriculture
800 Buchanan Street
Albany, CA 94710

Chapter 1

Examining Risks and Benefits Associated with Pesticide Use: An Overview

James N. Seiber [1] and Nancy N. Ragsdale [2]

[1] Western Regional Research Center, Agricultural Research Service, U. S. Department of Agriculture, 800 Buchanan Street, Albany, CA 94710
[2] Agricultural Research Service, U. S. Department of Agriculture, Building 005, BARC-W, Beltsville, MD 20705

Pesticides can improve the quality of our lives in and around our homes, in recreational and aesthetic areas, and, most importantly, in the year-around availability of agricultural produce, such as fruits and vegetables, which add diversity and nutritional quality to our diets. They have also been of great value in combating pests which transmit disease or otherwise adversely impact human health and are important tools in controlling the spread of pests imported from other countries. Pesticides have played a major role in improving agricultural production in its constant struggle to provide an adequate supply of food to mankind. The association of risks with pesticides has probably been recognized since pesticides were first used. The extent and nature of pesticide risks are much better documented and understood now than in the past, partly because of the development of toxicology as a discipline and the impressive gains in analytical chemistry, both relatively recent occurrences. However, neither benefits nor the process to determine them have ever been adequately defined. This paper will examine various risk and benefit factors that currently contribute to views affecting the availability of pesticides.

Pesticides have allowed twentieth century farmers to dramatically increase yields using less labor, less land, and less location-to-location, season-to-season, and year-to-year variations in yield, quality and cost. Modern agriculture makes use of extensive monoculture or oligoculture systems in order to maximize efficiency in producing, processing, and marketing food commodities. However, these systems are often more subject to weed, insect, fungus, rodent, virus and other pest infestations than the less efficient agricultural systems of the past, so that pest management becomes a key element in successful modern production (*1*).

The use of pesticides carries with it a variety of risks. Inherent toxicity and analytically measurable exposure are critical ingredients in assessing risks due to chemicals in general, including pesticides. For ethyl parathion, lead arsenate, dichlorodiphenyltrichloroethane (DDT), dibromochloropropane (DBCP), and several other chemicals, the risks have been judged by regulatory authorities and society to outweigh the benefits, resulting in voluntary or imposed cancellations or bans. Some states, such as California, have enacted legislation which imposes detailed scrutiny of pesticide risks from actual or potential human exposures in the workplace, and in air and drinking water. Increasingly, society has trended toward focusing more on the risks of pesticides than on benefits, sometimes, some would argue, appearing to conjure up risks (*2*) which may be trivial, scientifically controversial, or simply illogical.

The risks versus benefits arguments are usually adversarial to the extent that the real purpose of risk-benefit assessment, which is to stimulate continual improvement in pesticide agents and their human health and environmental safety, is lost. Risk-benefit assessments will promote continued reductions in risks and increased benefits to producers as well as consumers. However, the risk process must constantly be updated to reflect current scientific data, and the benefits data must be more robust. In order to continue the progress that has occurred in pest management over the last thirty years, science must play a stronger role in the decisions that determine what pesticides are available for use in agricultural production.

Food Quality Protection Act (FQPA)

The Food Quality Protection Act (FQPA), passed unanimously by Congress August 3, 1996, provided important amendments to both the Federal Insecticide, Fungicide, and Rodenticide Act (FIFRA) and the Federal Food Drug and Cosmetic Act (FFDCA), the two major laws governing the use of pesticides in the U.S. FQPA addressed a number of issues that had been in several bills before Congress in previous years. In addition, a 1993 National Academy of Sciences study, "Pesticides in the Diets of Infants and Children" (*3*), was influential in pointing out the need to address health risk issues from the perspective of infants and children. The FQPA amendments to FFDCA eliminated the application to pesticides of the Delaney Clause, which prohibits any food additive that has been shown to cause cancer in humans or laboratory animals. This does not mean that cancer risk will not be considered; it simply means that the parameters of the controversial Delaney Clause, which apply to processed commodities, will not apply in regulatory decisions. Key additions to the previous risk assessment process under FIFRA and FFDCA that result from FQPA include special consideration of infant/child exposure, determination of the aggregate risk that results from all exposures to a given chemical, determination of the cumulative risk that results from exposures to all chemicals with a common mechanism of toxicity, and assessment of information regarding the potential for pesticides and other chemicals to disrupt the human endocrine system.

Included among the many provisions of FQPA is an accelerated review and re-registration process for all pest-control chemicals, and a thrust for registering/re-registering reduced-risk and minor use pesticides. FQPA will drive many changes in the registration of pesticides, with implications for the discovery and development phases, use patterns, and safety evaluation. The full impacts of this legislation are yet to be felt, and, in fact, even in 1999 the U.S. Environmental Protection Agency has continued to struggle with the implementation of FQPA (*4*). A great deal of concern exists in the agricultural community about the continued availability of adequate pest management tools that are essential to the future success of U.S. agriculture.

Risks

Major strides have been made in improving the efficacy and safety of pesticides in the past thirty years, in part stimulated by Rachel Carson's "Silent Spring" (*5*), published in 1962, and the report of the Mrak Commission in 1969 (*6*). These improvements have resulted from greatly increased resources devoted to pesticide development and stewardship by industry, agencies, and academic institutions (*7*). Among the examples of improvement are the following:

- Pesticide-caused occupational and accidental mortality and illness has declined substantially in the U.S. (*8*).
- Formulation, application, and waste management practices have improved considerably (*9*).
- Overall market share of lower dose, less toxic, and less persistent pesticides has increased relative to the older, higher dose, more toxic, and more persistent chemicals (*10*).
- Food residue monitoring programs have consistently shown low or immeasurable residues, and generally low exposure for pesticides used in accordance with label instructions (*11*).

These areas of improvement demonstrate that, through scientific research and application, risks can be minimized. New types of pesticides, improved systems to ensure they reach intended target sites, assessment of actual human exposure, and bioremediation are all active research areas that will promote further risk reduction. As efforts continue to make improvements, the question has been raised about the role of the public sector in the research that underpins such improvements. The U.S. Department of Agriculture requested a study by the National Academy of Sciences, National Research Council, Board on Agriculture to assess and develop a report that will do the following:

- Identify the circumstances under which chemical pesticides may be required in future pest management.
- Determine what types of chemical products are the most appropriate tools for ecologically based pest management.
- Explore the most promising opportunities to increase the benefits, and reduce health and environmental risks of pesticide use.
- Recommend an appropriate role for the public sector in research, product development, product testing and registration, implementation of pesticide use strategies, and public education about pesticides.

Of course, research alone cannot minimize risks. Education and communication are key factors as well. The National Academy of Sciences report will be completed and made public in 1999.

Benefits

The benefits of pesticides, such as those in human health resulting from availability of a diverse, wholesome, year-round food supply at prices affordable by all segments of U.S. society, have not weighed as heavily as risks in recent societal evaluation of pest control agents. This is reflected in U.S. pesticide law. Under FIFRA, benefits were included in regulatory decisions; under the FQPA, benefits considerations have considerately diminished (*12*).

Failure to adequately consider benefits may be related to a number of factors. For one thing, no precise definition of benefits has ever been publicly accepted. The most obvious benefit, and that used most frequently in regulatory debates, is the economic advantage that agricultural producers receive. The second frequently cited benefit is the advantage that is passed to consumers providing access to quality agricultural commodities at reasonable prices. There are other considerations that should fit into the determination of benefits. These include such things as the impacts of production on the various industries (packaging, processing, transporting, retailing, etc.) associated with moving raw products to the consumer, and the health benefits that result from control of human disease vectors and from food sources free of toxins associated with pest infestations.

The issue of benefits goes back to the period following the 1975 amendment to FIFRA. This is the legislation that called for consideration of benefits in the regulatory decision making process. At this time benefits should have been defined, the methods to determine them outlined, and the process subjected to public comment followed by publication of an accepted procedure that would play an integral role in balancing benefits and risks. In the late 1970s the agricultural community was not sufficiently organized to accomplish this. In contrast, risks have been well defined, and methods for risk determination have been meticulously laid out, despite questions on the scientific relevance of the approach, going through public scrutiny by publication in the "Federal Register" and presentation to EPA's FIFRA Scientific Advisory Panel in meetings open to the public. Nothing like this has ever been attempted for benefits, and the vagueness surrounding the issue has caused numerous difficulties over the years, especially when specific pesticides which have been integral parts of agricultural production systems are subject to the possibility of cancellation. Determining benefits is not an easy task; there is a large number of variables in agricultural production. However, the agricultural community needs to be in the position to clearly lay out the likely impacts of various pest management scenarios that would serve as alternatives when chemical components of current practices are in question. To do this, the task of developing the methodology required to present a case that can withstand scrutiny must be undertaken.

In examining the benefits of a variety of pest management tools, the importance of alternatives to chemical pesticides and alternative production systems which rely less on chemicals for pest control must be recognized. Many such alternatives have emerged during the time period of 1970 to present. Production-scale demonstrations of non-chemical pest control systems are more frequent (*13*), and the rapid commercial development of agricultural biotechnology (*14*) promises whole new dimensions in agricultural pest control. However, improvements in chemicals (novel synthetics, natural products, identification of new targets and modes of action) are also accelerating, making even less likely an "either-or" scenario (*10*).

Part of the stimulus to move to alternative pest management systems is the desire to minimize or eliminate use of pesticide chemicals because of a perception that they are unsafe at any level of use or residue load (*15*). Another related stimulus is from the public's ready and steadily increasing acceptance of organic foods, which are assumed to be grown without the use of pesticides. However, another is purely economic; pesticide chemicals alone have not been able to solve all pest control problems, or can do so only at an unacceptable financial cost due to pest resistance and pest resurgence (*16*). Successful production systems must use an integrated pest management approach, taking advantage of the wide variety of avenues which minimize risks while increasing benefits.

Global Market

The ever-increasing amount of global trade has emphasized the need for harmonization of pesticide regulations. In fiscal year 1996 the U.S. exported agricultural, fish and wood products worth approximately $69.7 billion while importing such products worth approximately $49.8 billion (*17*). The FQPA encourages support of international harmonization efforts. Moving agricultural products grown in one country and intended for food use in another country raises questions about permissible residue levels, which of course are based on registration requirements. These, in turn, are primarily based on health and environmental factors. Common regulatory approaches could greatly simplify the current situation that results from wide variations in registration requirements from one nation to another.

Another factor that must be considered in global trade is exclusion of pests from environments in which they do not currently exist. For this purpose, pesticides are commonly used, and countries often require that certain treatments occur before accepting a shipment. Currently there is a great deal of concern about the fumigant, methyl bromide, which is often used in import/export of a wide variety of commodities to remove the possibility of pest infestations. In accordance with the Montreal Protocol, methyl bromide, as a likely stratospheric ozone layer depleter, is scheduled for phase out by 2005. Research is looking for alternatives that will permit continuation of global trade without increased risk of foreign pest introduction.

Agricultural production efficiency is improving rapidly in many developing countries. Pesticides have played an important role as these countries become more self-sufficient in food production, and, in some cases, even export agricultural produce. However, education is critical to assure that pesticides are stored, used and disposed of in a manner which will minimize health and environmental risks.

Conclusions

There is an underlying assumption that chemicals will continue their important use in agricultural pest management, at least until non-chemical alternatives are tested, ready, and economically competitive with chemicals. Most authorities predict that the sole use of non-chemical alternatives will not occur in the foreseeable (30-50 year horizon) future, if it occurs at all. The passage of FQPA will serve as a reminder of the continual, and even increasing, scrutiny which pesticides attract. This reinforces the need to better assess both risks and benefits, while constantly considering ever-changing societal criteria, including safety and economic concerns.

Literature Cited

1. Klassen, W. In *Eighth International Congress of Pesticide Chemistry - Options 2000*; Ragsdale, N. N.; Kearney, P. C.; and Plimmer, Eds.; American Chemical Society: Washington, DC, 1995; pp. 1-32.
2. Winter, C. K. *Food Technol.* **1998**, *52,*148.
3. *Pesticides in the Diets of Infants and Children.* National Research Council, National Academy Press; Washington, DC, 1993.
4. Cooney, C. M. *Environ. Sci. Technol.* **1999**, 33(1), 8A-9A.
5. Carson, R.L. *Silent Spring;* Houghton Mifflin Company: Boston, MA.1962.
6. *Report of the Secretary's Commission on Pesticides and their Relationship to Environmental Health.* U.S. Department of Health, Education, and Welfare,U.S. Government Printing Office: Washington, D.C., 1969.
7. Young, A. L. In *Pesticides: Minimizing the Risks*; Ragsdale, N. N.; Kuhr, R. J., Eds.; American Chemical Society: Washington, DC, 1987; pp. 1-11.
8. *California Pesticide Illness Surveillance Program Summary Reports for 1993,* 1994, and 1995. California Environmental Protection Agency, Department of Pesticide Regulation: Sacramento, CA, 1995, 1996, 1997.
9. *Pesticide Waste Management: Technology and Regulation;* Bourke, J. R.; Felsot, A. S.; Gilding, T. J.; Jensen, J. K.; Seiber, J. N., Eds.; American Chemical Society: Washington, DC, 1992.
10. *Pesticides and the Future*; Kuhr, R. J.; Motoyama, N., Eds.; IOS Press: Amsterdam, 1998.
11. *Pesticide Data Program, Annual Summary Calendar Year 1997*; Agricultural Marketing Service, USDA, Washington, DC, 1998.
12. Mintzer, E. S.; Osteen, C. *FoodReview* **1997,** *20, issue 1,* 18-26.
13. *Alternative Agriculture.* National Research Council, National Academy Press; Washington, DC, 1989.
14. Thayer, A.M. *Chem. Engr. News* **1997**, 75(17), 15-19.
15. Green, M.B. *Pesticides – Boon or Bane?* Westview Press: Boulder, CO, 1976.
16. *Pesticide Resistance: Strategies and Tactics for Management.* National Research Council, National Academy Press: Washington, DC, 1986.
17. *Agriculture Fact Book 1997.* Office of Communications, USDA, U. S. Government Printing Office: Washington, DC, 1997; pp. 92-96.

FOOD QUALITY PROTECTION ACT

Chapter 2

Food Quality Protection Act of 1996

Major Changes to the Federal Food, Drug, and Cosmetic Act; the Federal Insecticide, Fungicide, and Rodenticide Act and Impacts of the Changes to Pesticide Regulatory Decisions

Stephen L. Johnson and Joseph E. Bailey

Office of Pesticide Programs, U. S. Environmental Protection Agency, 401 M Street, S.W. (7501C), Washington, DC 20460

Disclaimer: The views presented in this paper are those of the authors and not necessarily those of the U.S. Environmental Protection Agency (EPA or the Agency).

The Food Quality Protection Act (FQPA) imposed new requirements on pesticide regulation by amending the Federal Insecticide, Fungicide and Rodenticide Act (FIFRA), and the Federal Food, Drug and Cosmetic Act (FFDCA). The FQPA established a new safety standard for pesticide residues in food -- ensuring a "reasonable certainty of no harm," with special consideration given to assessing potential risks to infants and children by including an additional ten-fold safety factor unless the Agency determines that a different factor is adequate. EPA's risk assessment process also changed to consider additive effects of pesticide exposure from multiple sources (e.g., drinking water, residential and dietary) and cumulative effects of pesticides which share common mechanisms of toxicity. EPA is required to reassess all tolerances, or allowable food residues, according to the new safety standard and to establish a plan to reevaluate registered pesticides periodically. EPA is required to establish an endocrine disruptor screening and testing program to identify pesticides that may affect endocrine processes. EPA no longer considers the "de minimis" risk standard, or the "Delaney Paradox," that the FFDCA required in establishing tolerances for pesticides classified as carcinogens. As a result of the FQPA requirements, pesticide regulatory decisions have been broadly impacted with a heightened food safety awareness.

The scale of agricultural production today is dependent upon the availability of pesticides in order to produce the quantity and quality of food demanded by the world's growing population. The Environmental Protection Agency (EPA) regulates the sale, distribution and use of pesticides under two statutes; the Federal Insecticide, Fungicide and Rodenticide Act (FIFRA), which gives the Agency the authority to register and label pesticides for use in the United States in such a way that they will not cause unreasonable adverse effects to human health or the environment; and the Federal Food, Drug and Cosmetic Act (FFDCA), under which the EPA establishes tolerances, or maximum legally permissible levels, for pesticide residues in or on food commodities.

On August 3, 1996, President Clinton signed into law the Food Quality Protection Act (FQPA) of 1996 making sweeping changes to the way the EPA regulates pesticides *(1)*. The FQPA, which was supported by the Administration and a broadly represented coalition of environmental, public health, agricultural and industry groups, is the first major revision of laws governing pesticide regulation in over 30 years and, it was unanimously signed into law by both houses of Congress with little resistance. The FQPA greatly strengthens the regulations which protect the Nation's food supply from potentially unsafe pesticide residues and sets more strict standards that must be met in order to satisfy registration requirements for pesticides used on food commodities. Of particular interest, is the emphasis that this legislation places upon the protection of infants and children from exposure to pesticides due to their increased sensitivity as compared with adults.

When the FQPA was signed into law, the EPA was faced with the daunting task of implementing the law without any allowance provided for a phase-in period. Risk assessment processes needed to be changed significantly to accommodate the requirements of the new law; committees needed to be established to address issues such as identifying pesticides that affect endocrine functioning; and a mechanism needed to be put into place to reassess within 10 years almost 10,000 tolerances according to a new safety standard. The EPA's already heavy burden of regulating pesticides was only made more burdensome by the passage of the FQPA; however, the importance that the law places upon protecting public health and ensuring the safety of the Nation's food supply far outweigh the cost of any additional resource demands that the law has required of the EPA.

Amendments to the Federal Food, Drug and Cosmetic Act

Although the primary statute under which the EPA registers, and therefore regulates pesticides, is FIFRA, the predominant effect that the FQPA has had on the pesticide regulatory process is through the amendments made to the FFDCA. It is under the authority of the FFDCA that the EPA establishes tolerances for residues of pesticides in or on food commodities, and it is in this area, that the FQPA has made the most profound changes. The regulation of pesticides through the tolerance process is shared by the EPA and the Food and Drug Administration (FDA). While it is the EPA's responsibility to determine what level of pesticide residue can be allowed to remain in or on foods from a health standpoint, it is the FDA's responsibility to monitor food items to ensure that the legally enforceable levels established by the EPA are not exceeded.

A significant improvement to pesticide legislation brought about by the FQPA was repeal of the Delaney Clause as it relates to pesticide residues and tolerances. Previously, pesticide residues in processed foods were considered to be food additives and, if residues in the processed food exceeded the FFDCA section 408 tolerance for the raw commodity, a separate tolerance was required under section 409 for the processed food *(2)*. In addition, strict interpretation of the Delaney Clause allowed a zero level of residues in food commodities of pesticides that were classified as carcinogens. Under the new law, pesticide residues are not considered to be food additives and therefore, are no longer subject to regulation under the Delaney Clause*(3)*. Instead, the FQPA has established a single health-based standard which applies to both raw agricultural commodities as well as processed foods---a standard that is much less contradictory and easier to apply to pesticide regulatory policy. The single health-based standard is such that the EPA can establish or maintain a tolerance if it is determined to be safe, and safe is defined in the law to mean that there is a reasonable certainty that no harm will result from exposure to the pesticide residue*(4)*.

The FQPA emphasizes the importance of considering extra sensitivities of infants and children to pesticide exposure. In establishing, modifying or revoking a tolerance, the EPA must now consider in its risk assessment process, any available information about food consumption by infants and children, information about increased sensitivity, and information about cumulative effects of pesticides that may share a common toxic mechanism. The 1993 National Academy of Sciences study, "Pesticides in the Diets of Infants and Children," reported the results of its research on what is known about the effects of pesticides in the diets of infants and children and evaluated current risk assessment methodologies and toxicological issues of concern*(5)*. In general, the report concluded that infants and children may react differently from adults when exposed to pesticides and that these differences should be considered when assessing potential risks from exposure. Further, the report recommended that, in instances where increased susceptibility to pesticides is believed to occur, an additional safety factor should be more routinely employed to adequately protect infants and children from potential risks. It is from some of the recommendations of this report that the FQPA mandates were framed into legislative initiatives for pesticide regulatory policy reform, particularly focusing on the protection of infants and children.

Stemming from the recommendations of the National Academy of Sciences report, the FQPA requires that up to an extra 10-fold safety factor be applied during risk assessing when it is determined that a pesticide may present risks to infants and children because of their increased sensitivity*(6)*. As a result, the EPA places greater emphasis on its review of toxicological studies that provide insights to reproductive, developmental and neurological effects of pesticides that could indicate increased susceptibility to infants and children. In those cases where extra sensitivities are believed to be possible, after using a weight-of-evidence evaluation, the Agency will retain an additional 10-fold safety factor in order to adequately protect these more sensitive individuals unless there are reliable data that indicate a lower safety factor, or no additional factor, is adequately protective. A standard 100-fold safety factor has always been used by the EPA in its risk assessments to account for intra- and inter-species variability and uncertainty.

The EPA has historically handled each chemical and each exposure scenario separately in its risk assessment determinations. Risks have been estimated for diet, drinking water, and residential uses, such as pest control in the home or on lawns and in gardens, but they were not added together to see what the combined risk estimate would be. The FQPA has addressed the fact that people are not exposed to chemicals on an individual basis, but rather, may be subject to exposure to several different chemicals simultaneously, and from a variety of sources. Therefore, it is reasonable to approach risk assessment in such a way that reflects actual exposure in the real world, and the

FQPA now requires that the EPA consider this issue through both aggregate exposure and through cumulative risk assessment*(7)*. The EPA is required to determine that a reasonable certainty of no harm will result when considering the additive effects of various exposure routes for a single pesticide; i.e., exposure from dietary sources (including food and drinking water) and all other non-occupational sources, largely residential exposure. Similarly, the EPA is required to consider the additive or cumulative effects of those pesticides which exhibit toxicological effects through similar mechanisms of action.

At the time the FQPA was signed by the President, the EPA had about 9,700 tolerances on record. The FQPA requires that all existing tolerances be reviewed according to the new safety standard established by the legislation and an ambitious schedule for review of these tolerances is also established by the law. The FQPA requires that 33 percent of the tolerances on record be reviewed within 3 years, 66 percent within 6 years and 100 percent within 10 years*(8)*. In August 1997, the EPA published a notice in the *Federal Register* that outlined a schedule to meet this requirement*(9)*. Also, the law requires that the EPA give priority to reassessing the tolerances for those chemicals which appear to pose the greatest risk. The EPA is advancing review of those pesticides that pose the highest risks to the front of the queue and is developing an approach to review the organophosphate pesticides as the first major group of similar chemicals. Sharing a common endpoint, cholinesterase inhibition, it has been recommended that risk assessment for this class of pesticides be conducted cumulatively as the FQPA requires for chemicals sharing a common mechanism of toxicity.

The FQPA further requires the EPA to establish a tolerance reassessment fee system that adequately maintains the services required to reassess tolerances, including the acceptance for filing of a petition and for establishing, modifying, leaving in effect, or revoking a tolerance or exemption from the requirement for a tolerance*(10)*. The funds generated shall be available without fiscal year limitations.

The FQPA has changed how benefits may be considered in determining the eligibility of a pesticide for registration. The new law allows tolerances to remain in effect for pesticides that might not otherwise meet the new safety standard based on benefits of the pesticide only under certain conditions. Pesticide residues would only be "eligible" for such tolerances if use of the pesticide prevents even greater health risks to consumers or the lack of the pesticide would result in "a significant disruption in domestic production of an adequate, wholesome and economical food supply*(11)*." The new provision narrows the range of circumstances in which benefits consideration plays a significant role in determining whether or not a tolerance is appropriate for a particular pesticide use.

The FQPA emphasizes increased public awareness about the foods we eat and the potential for pesticide residues to occur on those foods and mandates the EPA to publish by August 1998, a document that discusses the risks and benefits of pesticides, a list of those pesticides for which there are benefit-based tolerances and the foods which may contain residues of these pesticides, and recommendations for ways to reduce dietary exposure to residues of pesticides in foods*(12)*. The document is required to be provided to major grocery stores and made available on an annual basis after the first publication. Grocers may decide how they wish to make the information available, and in fact, are not bound in any legal manner to even make the information available to consumers.

However, it is a means of letting consumers know how to reduce potential risks that might be possible from pesticide residues in food and allows them to make more informed decisions about the kinds of foods they eat. The EPA fully endorses measures to inform the public about all aspects about the potential risks and benefits of pesticides.

Endocrine disruptors are chemicals which are believed to conflict with normal functioning of natural endocrine hormones in animals and humans. Although there is little, if any, demonstrative evidence of disruptive effects in humans, data are available which indicate that certain chemicals are biologically active in affecting hormonal functions in certain wildlife, and therefore, it is not unreasonable to believe that some effects may be possible in humans. This area of science is relatively newly emerging and much is unknown about the actual effects; however, continuing research is elucidating some of the science. The FQPA acknowledges the potential effects on endocrine functioning that may be linked with pesticide exposure and requires the EPA to develop a screening and testing program for pesticides that will determine if certain chemicals may have endocrine disrupting effects*(13)*.

Amendments to the Federal Insecticide, Fungicide and Rodenticide Act

The Federal Insecticide, Fungicide and Rodenticide Act was first enacted in 1947 to regulate the use of pesticides in the United States with a number of amendments having been made subsequently to the law*(14)*. Pesticide use is largely controlled by the registration or approval of specific uses of pesticides by the EPA based on scientific review of data about the specific chemical and a risk assessment/risk management process that determines the conditions under which the pesticide may or may not be used. The FQPA changes to FIFRA relate largely to minor uses and antimicrobial registration.

Minor crop consideration by industry has always been a concern for the United States Department of Agriculture (USDA), the EPA and minor crop growers. Because of lower economic incentives to produce chemicals for minor crop production, industry has tended not to be as supportive of chemicals for minor uses as compared to other larger scale production crops. The FQPA first clearly defines minor use as use of a pesticide on a crop that has less than 300,000 total U.S. acres in production. The FQPA then provides incentives that make supporting minor use pesticides more appealing. These incentives generally focus on the submission of data to support minor use pesticide registration, such as providing time extensions for providing required data and expediting review of applications for minor uses. In addition, the EPA is required by the FQPA to establish a minor use program to coordinate minor use activities*(15)*. In September 1997, the Agency formed a minor use team designed to provide a coordinated program-wide approach to minor use pesticide issues. The goals of the team are three-fold: 1) to promote the collection and use of best available usage information for risk assessments, 2) to facilitate open dialogue with the minor use community and 3) to promote the development of safer pesticides for minor use crops. The EPA is committed to working closely with the USDA and growers to ensure that needed pesticides continue to be available to control minor use pests.

Prior to the FQPA, no special provisions were in place for the regulation of antimicrobial pesticides and the EPA provided no special consideration for antimicrobial pesticide applications for registration. The new law requires reform of the antimicrobial registration review process to accelerate reviews for registration of antimicrobial pesticides and their amendments*(16)*. Since the FQPA was enacted, the Agency has structured a separate division solely devoted to the registration related actions for antimicrobial pesticides. Focused attention has been placed on the number of pending actions for antimicrobials and as a result, backlogs have been reduced by more than 75 percent, reflecting a diligent effort to provide faster processing under a more streamlined process.

Perhaps one fundamental change to pesticide regulatory policy that the FQPA imposes is acknowledgment that science is not a static discipline and pesticide research and risk assessment methodology is a rapidly developing area. The EPA has undertaken reviews of pesticides previously registered under FIFRA in discrete programs, most recently under the 1988 amendments to FIFRA that led to establishment of a program to review all pesticide active ingredients registered prior to November 1984. The intent of this reregistration program is to bring those active ingredients registered prior to November 1984 up to current day standards of testing as required by the EPA, but only on a one-time basis, with no further reviews required. Because science is continually evolving, the FQPA requires the EPA to periodically review pesticide registrations with a review goal of every 15 years*(17)*. The EPA is required to establish, by regulation, a procedure for this periodic review. If the EPA determines that additional data are needed for any review, such data may be required under FIFRA section 3(c)(2)(B)*(18)*. This periodic revisiting of pesticide registrations ensures that their physicochemical and toxicological characteristics are reevaluated according to current state-of-the-science.

EPA's Direction and Impacts of the FQPA on Regulatory Decisions

The enactment of the FQPA has presented challenges to everyone involved---growers, regulators, industry, environmental groups and government alike. While the EPA has taken steps to implement all of the requirements of the law, those requirements that emphasize the protection of infants and children have perhaps taken front stage as deliberations progress toward policy development and refinement. Those aspects of the FQPA that require consideration of an additional safety factor, aggregate exposure and cumulative risk are difficult issues to resolve and are complex science issues at the leading edge of evolving pesticide toxicology and risk assessment methodology. The Agency is working to develop sound science policy for these issues and is consulting experts in these areas to allow the best scientific minds to contribute to the resolution of difficult problems.

Because the requirements of the FQPA could impact a wide range of regulators and stakeholders, it has been the EPA's intent to involve as many of the groups that may be affected as early as possible in the implementation stage. This is particularly the case with the difficult science issues that the Agency needs to understand before taking regulatory action based on new policy. Such issues include how to apply the additional 10-fold safety factor, how to incorporate aggregate exposure and cumulative risks into the risk assessment/risk management process, how to screen for endocrine disrupting pesticides, how to incorporate drinking water and residential exposure assessments into the overall risk assessment for the chemical, and how to design an effective consumer right-to-know document that provides information about lowering potential risks from pesticide residues in food. Through such forums as the FIFRA Scientific Advisory Panel; the Pesticide Program Dialogue Committee; and the International Life Sciences Institute, a non-profit worldwide foundation established to advance the understanding of scientific issues related to nutrition, food safety, toxicology and the environment; the Agency is providing opportunity for public involvement in our policy making process. In March, the Agency presented several major issues to the Scientific Advisory Panel that included consideration of the common mechanism of action for the organophosphate pesticides, possible probabilistic risk assessment methodology for evaluating pesticides that exhibit common mechanisms of action, and the use of the additional 10-fold safety factor to address special sensitivity of infants and children to pesticides. Elucidating the science around such complex issues requires the knowledge of the best experts in the field to adequately establish sound science policy and regulatory decisions. It is the EPA's goal to do this. Further, the Agency is urging growers to communicate with the EPA through its representative organizations and through the USDA to let the EPA know more precisely how particular pesticides are used in their cropping schemes and what alternative pesticides they will resort to if certain pesticide uses are determined to be ineligible for continued use under the new FQPA safety standard.

In response to an April 8, 1998 memorandum from Vice President Al Gore to the EPA and the USDA in which the Vice President reaffirmed the Administration's commitment to the FQPA and clarified how to fulfill the requirements of the law *(19)*, an advisory group was created to ensure smooth implementation of the requirements so that the important health aspects of the law are carried out and at the same time, the Nation's important agricultural production is not impeded. The advisory group, co-chaired by the Deputy Administrator of the EPA and the Deputy Secretary of the USDA, will ensure that the implementation of regulatory processes flowing from the FQPA requirements is transparent, based on sound science, and provides for a reasonable transition for agriculture that reduces risk from pesticide use while not jeopardizing the level of agricultural food production. An unprecedented level of consultation will occur between the EPA and the USDA as well as the public, other Federal agencies, including the FDA and the Center for Disease Control, and other qualified participants representing farmers, pesticide companies, environmental groups, public interest groups, and state, tribal, and local governments. The advisory group will be requested to lay a framework to review the first major class of chemicals the EPA is evaluating, the organophosphates.

About a year and a half has passed since the FQPA was enacted and the EPA has continued to make regulatory decisions for pesticide actions throughout this period, despite some impacts of immediate implementation. Overall, the number of decisions the EPA made during the first year were slightly under previous records; however, not significantly. The nature of some of the decisions have been affected by the new safety standard imposed by the FQPA. Some uses, particularly for emergency exemptions under FIFRA section 18, have been denied because of the inability to make a reasonable certainty of no harm finding. The EPA does expect that more difficult decisions will need to be made in the future as reviews of pesticides such as the organophosphates are completed and aggregate and cumulative considerations for classes of compounds are dealt with. However, the EPA will not make these difficult decisions in isolation and fully intends to keep all interested parties informed of the processes that will be used to evaluate chemicals and what specific chemical decisions are under consideration.

Conclusion

The EPA has taken significant steps to implement all of the provisions of the FQPA. The EPA's goal has been to provide opportunities for as much public participation in developing FQPA policy as is possible, while not significantly impacting the number of decisions we are expected to make in order to carry out our mission of protecting public health and the environment. The Agency believes that involving stakeholders and other interested parties in developing new policies is better as a whole and the decisions that will ultimately result from such policy will be better, more informed decisions. The Agency is committed to protecting public health and the environment and has placed a heightened awareness around the special sensitivities of infants and children and towards strengthening programs that improve consumers' right-to-know about pesticides and the safe, responsible use of them. As the Agency moves forward with the implementation of the FQPA, efforts will be made to continue to keep the regulated community, the environmental community, consumers, other governmental agencies, and growers aware of the complex issues we are dealing with and our approaches to find answers to the difficult questions that are certain to arise.

Literature Cited

1. *Food Quality Protection Act of 1996*, Public Law 104-170, 1996.
2. *Federal Food Drug and Cosmetic Act*, 21 U.S.C., 1994.
3. *Federal Food Drug and Cosmetic Act*, § 201(s).
4. *Federal Food Drug and Cosmetic Act*, § 408(b)(2)(A).
5. National Research Council. *Pesticides in the Diets of Infants and Children*; National Academy Press: Washington, DC, 1993; p 2.
6. *Federal Food Drug and Cosmetic Act*, § 408(b)(2)(C), 1996.
7. *Federal Food Drug and Cosmetic Act*, § 408(b)(2)(D), 1996.
8. *Federal Food Drug and Cosmetic Act*, § 408(q), 1996.
9. Federal Register Volume 62, Number 149, pgs 42019-42030, August 4, 1997.
10. *Federal Food Drug and Cosmetic Act*, § 408(m), 1996.
11. *Federal Food Drug and Cosmetic Act*, § 408(b)(2)(B), 1996.
12. *Federal Food Drug and Cosmetic Act*, § 408(o), 1996.
13. *Federal Food Drug and Cosmetic Act*, § 408(p), 1996.
14. *Federal Insecticide, Fungicide and Rodenticide Act*, Public Law 100-460, 100-464 to 100-526, and 100-532, 1988.
15. *Federal Insecticide, Fungicide and Rodenticide Act*, § 31(a), 1996.
16. *Federal Insecticide, Fungicide and Rodenticide Act*, § 3(h)(2), 1996.
17. *Federal Insecticide, Fungicide and Rodenticide Act*, § 3(g)(1)(A), 1996.
18. *Federal Insecticide, Fungicide and Rodenticide Act*, § 3(c)(2)(b), 1996.
19. Memorandum from the Vice President to Secretary Daniel R. Glickman (USDA) and Administrator Carol M. Browner (EPA), April 8, 1998.

Chapter 3

Evaluating Exposures of Infants and Children to Pesticides

R. D. Thomas

INTERCET, Ltd.
International Center for Environmental Technology,
1307 Dolley Madison Boulevard, McLean, VA 22101-3913

Pesticides are used widely in agriculture to increase crop yields and have resulted in significant increases in the quantity and quality of fresh fruits and vegetables in the diet, thereby contributing to improvements in public health. Even so, human exposure to pesticides may also cause harm, if doses are elevated. Depending on the dose, a range of adverse effects in humans may be observed, including both acute and chronic injury to the nervous system, lung damage, reproductive dysfunction, and possibly dysfunction of the endocrine and immune systems. For children, diet is an important potential source of exposure. Children may be exposed to multiple pesticides with a common toxic effect, and estimates of exposure and of risk should therefore account for simultaneous exposures in a variety of foods. This paper will describe approaches to estimating exposures in children.

Pesticides are widely used in agriculture in the United States. Their application has improved crop yields and increased the quantity of fresh fruits and vegetables in our diet, thereby contributing to improvements in public health. Pesticides may also produce harm. They may damage the environment and accumulate in ecosystems. Depending on dose, pesticides may cause a range of adverse effects on human health, including cancer, acute and chronic injury to the nervous system, lung damage, reproductive dysfunction, and possibly dysfunction of the endocrine and immune systems.

Diet is an important source of exposure to pesticides. The trace quantities of pesticides that are present on or in foodstuffs are termed residues. To minimize exposure of the general population to pesticide residues in food, the U.S. Government has instituted regulatory controls on pesticide use. These are intended to limit exposures to residues while ensuring an abundant and nutritious food supply.

The legislative framework for these controls was established by the Congress through the Federal Insecticide, Fungicide, and Rodenticide Act (FIFRA), the Federal Food, Drug, and Cosmetic Act (FFDCA) and more recently, the Food Quality Protection Act (FQPA). Pesticides are defined broadly in this context to include insecticides, herbicides, and fungicides.

Tolerances constitute the single, most important mechanism by which EPA limits levels of pesticide residues in foods. A tolerance is defined as the legal limit of a pesticide residue allowed in or on a raw agricultural commodity and, in appropriate cases, on processed foods. A tolerance must be established for any pesticide used on any food crop. Tolerance concentrations are based primarily on the results of field trials conducted by pesticide manufacturers and are designed to reflect the highest residue concentrations likely under normal conditions of agricultural use. Their principal purpose is to ensure compliance with good agricultural practice. Tolerances are not based primarily on health considerations.

Concern about the potential vulnerability of infants and children to dietary pesticides led the U.S. Congress in 1988 to request that the National Academy of Sciences (NAS) appoint a committee to study this issue through its National Research Council (NRC). In response, the NRC appointed a Committee on Pesticide Residues in the Diets of Infants and Children.

The Committee was charged with responsibility for examining the scientific and policy issues faced by government agencies, particularly EPA, in regulating pesticide residues in foods consumed by infants and children. Specifically, the committee was asked to examine the adequacy of current risk assessment policies and methods; to assess information on the dietary intakes of infants and children; to evaluate data on pesticide residues in the food supply; to identify toxicological issues of greatest concern; and to develop relevant research priorities. This presentation summarizes the results of the Committee's work and some of my own thoughts on assessing risk in children. A more detailed description of the Committee's work may be found in the NRC publication, "Pesticides in the Diets of Infants and Children" (1).

Age-Related Variation in Susceptibility and Toxicity

A fundamental principle of pediatric medicine is that children are not "little adults." Important differences exist between children and adults. Infants and children are growing and developing. Their metabolic rates are more rapid than those of adults. There are important differences in their ability to activate, detoxify, and excrete xenobiotic compounds. All these differences can affect the toxicity of pesticides in infants and children. Children may be more sensitive or less sensitive than adults, depending on the pesticide to which they are exposed. Moreover, because these processes can change rapidly with growth and can counteract one another, there is no simple way to predict the kinetics and sensitivity to chemical compounds in infants and children from data derived entirely from adult humans or from toxicity testing in adult or adolescent animals.

The committee found both quantitative and occasionally qualitative differences in toxicity of pesticides between children and adults. Qualitative differences in toxicity are the consequence of exposures during special windows of vulnerability. These are brief periods early in development when exposure to a toxicant can permanently alter the structure or function of an organ system. Classic examples include chloramphenicol exposure of newborns and vascular collapse (gray baby syndrome), tetracycline and dysplasia of the dental enamel, and lead and altered neurologic development.

Quantitative differences in pesticide toxicity between children and adults are due in part to age-related differences in absorption, metabolism, detoxification, and excretion of xenobiotic compounds, that is, to differences in both pharmacokinetic and pharmacodynamic processes. Differences in size, immaturity of biochemical and physiological functions in major body systems, and variation in body composition (water, fat, protein, and mineral content) all can influence the extent of toxicity. Because newborns are the group most different anatomically and physiologically from adults, they may exhibit the most pronounced quantitative differences in sensitivity to pesticides. In those studies examined by the Committee, they found that quantitative differences in toxicity between children and adults are usually less than a factor of approximately 10-fold.

Further they found that the mechanism of action of a toxicant, how it causes injury, is generally similar in most species and across age and developmental stages within species. For example, if a substance is cytotoxic in adults, it is usually also cytotoxic in immature individuals. However, the lack of data on pesticide toxicity in children was a recurrent problem encountered during the study. Little work has been done to identify the effects that develop after a long latent period or to investigate the effects of pesticide exposure on neurotoxic, immunotoxic, or endocrine responses in infants and children.

The Committee reviewed current EPA requirements for toxicity testing by pesticide manufacturers, as well as testing modifications proposed by the agency. In general, the committee found that current and past studies conducted by pesticide manufacturers are designed primarily to assess pesticide toxicity in sexually mature animals. Only a minority of testing protocols have supported extrapolation to infant and adolescent animals. Current testing protocols do not, for the most part, adequately address the toxicity and metabolism of pesticides in neonates and adolescent animals or the effects of exposure during early developmental stages and their sequelae in later life.

Age-Related Differences in Exposure

Estimation of the exposures of infants and children to pesticide residues requires information on (1) dietary composition and (2) residue concentrations in and on the food and water consumed. The committee found that infants and children differ both qualitatively and quantitatively from adults in their exposure to pesticide residues in foods. Children consume more calories of food per unit of body weight than do adults. But at the same time, infants and children consume far fewer types of foods than do adults. Thus, infants and young children may consume much more of certain foods, especially processed foods, than do adults. And water consumption, both as drinking water and as a food component, is very different between children and adults.

The Committee concluded that differences in diet and thus in dietary exposure to pesticide residues account for most of the differences in pesticide-related health risks that were found to exist between children and adults. Differences in exposure were generally a more important source of differences in risk than were age-related differences in toxicologic vulnerability.

Data from various food consumption surveys were examined during the study. They found it necessary to create their own computer programs to convert foods as consumed into their component raw agricultural commodities (RACs). This analytic approach facilitated the use of data from different sources and permitted evaluation of total exposure to pesticides in different food commodities. For processed foods, the Committee noted that effects of processing on residue concentrations should be considered, but that information on these effects is quite limited. Processing may decrease or increase pesticide residue concentrations.

The limited data available suggest that pesticide residues are generally reduced by processing; however, this remains an area where more research is needed to define the direction and magnitude of the changes for specific pesticide-food combinations. The effect of processing is an important consideration in assessing the dietary exposures of infants and young children, who consume large quantities of processed foods, such as fruit juices, baby food, milk, and infant formula.

Although there are several sources of data on pesticide residues in the United States, the data are of variable quality, and there are wide variations in sample selection, reflecting criteria developed for different sampling purposes, and in analytical procedures, indicating different laboratory capabilities and different levels of quantification between and within laboratories. These differences reflect variations in precision and in the accuracy of methods used and the different approaches to analytical issues, such as variations in limit of quantification. There also are substantial differences in data reporting. These differences are due in part to different record-keeping requirements, such as whether to identify samples with multiple residues, and differences in statistical treatment of laboratory results below the limit of quantification.

Both government and industry data on residue concentrations in foods reflect the current regulatory emphasis on average adult consumption patterns. The committee found that foods eaten by infants and children are underrepresented in surveys of commodity residues. Many of the available residue data were generated for targeted compliance purposes by the Food and Drug Administration (FDA) to find residue concentrations exceeding the legal tolerances established by the EPA under FFDCA. FQPA addresses some of these issues.

Survey data on consumption of particular foods are conventionally grouped by broad age categories. The average consumption of a hypothetical "normal" person is then used to represent the age group. However, in relying solely on the average as a measure of consumption, important information on the distribution of consumption patterns is lost. For example, the high levels of consumption within a particular age group are especially relevant when considering foods that might contain residues capable of causing acute toxic effects. Also, geographic, ethnic, and other differences may be overlooked.

To overcome the problems inherent in the current reliance on "average" exposures, the NRC used the technique of statistical convolution (i.e., combining various data bases) to merge distributions of food consumption with distributions of residue concentrations. This approach permits examination of the full range of pesticide exposures in the U.S. pediatric population. As is described in the next section, this approach provides an improved basis over the approach now used for assessing risks for infants and children.

New Approaches to Risk Assessment

To properly characterize risks to infants and children from pesticide residues in the diet, information is required on (1) food consumption patterns of infants and children, (2) concentrations of pesticide residues in foods consumed by infants and children, and (3) toxic effects of pesticides, especially effects that may be unique to infants and children. If suitable data on these three items are available, risk assessment methods based on the technique of statistical convolution and other related statistical approaches can be used to estimate the likelihood that children, based on specific exposure patterns, may be at risk. To characterize potential risks to infants and children, the NRC Committee utilized data on distributions of pesticide exposure. These were based on distributions of food consumption merged with data on the distribution of pesticide residue concentrations. Using this approach, the NRC found that age-related differences in exposure patterns for 1- to 5-year-old children were most accurately illuminated by using 1-year age groupings of data on childrens food consumption.

Exposure estimates should be constructed differently depending on whether acute or chronic effects are of concern. Average daily ingestion of pesticide residues is an appropriate measure of exposure for a the risk of chronic toxicity. However, actual individual daily ingestion is more appropriate for assessing acute toxicity. Because chronic toxicity is often related to long-term average exposure, the average daily exposure to pesticide residues may be used as the basis for risk assessment when the potential for delayed, irreversible chronic toxic effects exists. Because acute toxicity is more often mediated by peak exposures that occur within short time periods (e.g., over the course of a day or even from a single ingestion), individual daily intakes are important for assessing acute risks. Further, examining the distribution of individual daily intakes within a population of interest reflects the day-to-day variation in pesticide ingestion both for individuals and among individuals.

Children may be exposed to multiple pesticides with a common effect, and estimates of exposure and of risk could therefore be improved by accounting for these simultaneous exposures. One way this can be accomplished is by assigning toxicity equivalence factors to each compound with a common mechanism of action. Total residue exposure is then estimated by multiplying the actual level of each pesticide residue by its equivalence factor and adding the results. This information is then combined with data on consumption to construct a distribution of exposure to all pesticides having a common mechanism of action. Using this multiple-residue methodology, the committee estimated acute health risks resulting from combined exposure to five members of the organophosphate insecticide family (Figure 1).

Although some risk assessment methods take into account changes in exposure with age, these methods have not been universally applied in practice. The committee explored the use of newer risk assessment methods that allow for changes in exposure and susceptibility with age. However, the committee found that sufficient data are not currently available to permit wide application of these methods.

Given adequate data on food consumption and residues, the Committee recommended the use of exposure distributions rather than single point data to characterize the likelihood of exposure to different concentrations of pesticide residues. The distribution of average daily exposure of individuals in the population of interest is most relevant for use in chronic toxicity risk assessment, and the distribution of individual daily intakes is recommended for evaluating acute toxicity. Ultimately, the collection of suitable data on the distribution of exposures to pesticides will permit an assessment of the proportion of the population that may be at risk.

Although the Committee considers the use of exposure distributions to be more informative than point estimates of typical exposures, the data available to the committee did not always permit the distribution of exposures to be well characterized. Existing food consumption surveys generally involve relatively small numbers of infants and children, and food consumption data are collected for only a few days for each individual surveyed. Depending on the purpose for which they were originally collected, residue data may not reflect the actual distribution of pesticide residues in the food supply. Since residue data are not developed and reported in a consistent fashion, it is generally not possible to pool data sets derived from different surveys. Consequently, the Committee recommended that guidelines be developed for consumption and residue data permitting characterization of distributions of dietary exposure to pesticides.

For carcinogenic effects, the Committee proposed new methods of cancer risk assessment designed to take differences in susceptibility between children and adults into account. Preliminary analyses conducted by the committee suggested that consideration of such differences can lead to lifetime estimates of cancer risk that can be higher or lower than estimates derived with methods based on similar susceptibility and constant exposure. However, underestimation of risk assuming constant exposure was limited to a factor of about 3- to 5-fold in all cases considered by the Committee.

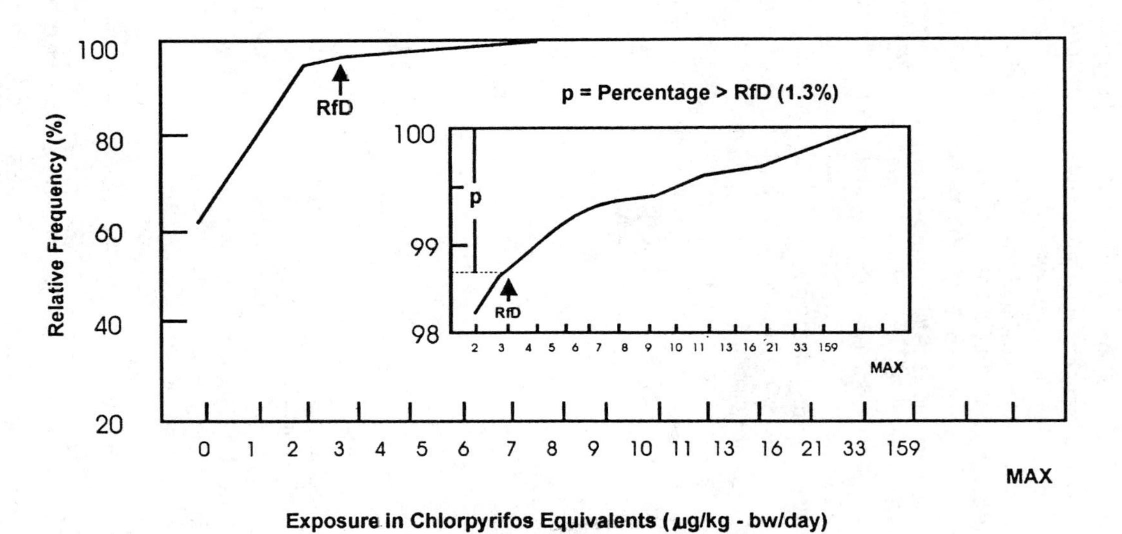

Figure 1. Exposure of 2-Year-Old Children to Organophosphate Pesticides. Intake Foods are Apples, Oranges, Grapes, Beans, Tomatoes, Lettuce, Peaches, and Peas. Chemicals are Acephate, Chlorphyrifos, Dimethoate, Disulfoton, and Ethion. The RfD is 0.003 mg/kg/bw/day. Modified from NRC, 1993.

Currently, most long-term laboratory studies of carcinogenesis and other chronic end points are based on protocols in which the level of exposure is held constant during the course of the study. To facilitate the application of risk assessment methods that allow for changes in exposure and susceptibility with age, it would be desirable to develop bioassay protocols that provide direct information on the relative contribution of exposures at different ages to lifetime risks. Although the Committee did consider it necessary to develop special bioassay protocols for application in the regulation of pesticides, it would be useful to design special studies to provide information on the relative effects of exposures at different ages on lifetime cancer and other risks with selected chemical carcinogens.

Conclusions

Better data on dietary exposure to pesticide residues should be combined with improved information on the potentially harmful effects of pesticides on infants and children. When assumptions are substituted for actual data, assumptions should be realistic, and the basis for these assumptions should be clearly stated. Risk assessment methods that enhance the ability to estimate the magnitude of these effects should be developed, along with appropriate toxicological tests for perinatal and childhood toxicity.

Literature Cited

1. Committee on Pesticides in the Diets of Infants and Children, *"Pesticides in the Diets of Infants and Children."* National Research Council, National Academy Press, Washington, D.C. , 1993.

Chapter 4

Issues Related to Screening and Testing for Endocrine Disrupting Chemicals

James C. Lamb, IV and Shanna M. Brown

Jellinek, Schwartz & Connolly, Inc., 1525 Wilson Boulevard, Suite 600, Arlington, VA 22209

Recently, the Environmental Protection Agency's (EPA) Endocrine Disruptor Screening and Testing Committee (EDSTAC) completed spelling out its recommendations for screening and testing. That report sets precedent for how the government could address the study and regulation of certain noncancer endpoints. We have learned many lessons from the use of short-term screening tests in cancer risk assessment. How can we apply those lessons to the environmental endocrine issue, when such screening is about to begin? How will we determine whether changes in male reproduction are real, and whether they are related to environmental chemical exposures? A good deal of research on breast cancer is underway, but significant questions exist in male reproductive health as well. These should receive significant attention. Which wildlife models are relevant to human health? It has been stated that the affected wildlife populations might serve as a sentinel for potential human effects. This issue should be addressed directly. The following presents an overview of these issues and describes how they may affect toxicology and risk assessment.

Endocrine disrupting chemicals (EDCs) have been the subject of many recent conferences and review publications. One of the first books to present endocrine disruption as an emerging issue was based on a "consensus conference" convened by Dr. Theo Colborn *(1)*. That meeting has often been referred to as the Wingspread conference. The proceedings of the conference were published, and they represent one of the first times the term "endocrine disruptor" was coined. According to the proceedings, the panel perceived a link between human outcomes and changes observed in wildlife populations. The consensus panel at that

conference expressed concern that, based on observations in wildlife populations, humans might be at risk for widespread adverse effects to their reproductive and endocrine systems from environmental chemicals. They expressed certainty that "a large number of man-made chemicals that have been released into the environment, as well as a few natural ones, have the potential to disrupt the endocrine system of animals, including humans" [that] "many wildlife populations are already affected" [and that] "humans have been affected by compounds of this nature, too." The Wingspread monograph made many predictions of adverse effects on reproduction and endocrine function caused by chemicals that mimic the effects of natural hormones through the natural ligand-receptor system *(1)*.
Shortly after the Wingspread conference, Dr. Colborn and her colleagues published a list of more than 40 chemicals ["known to affect the reproductive and endocrine systems"] *(2)*. That list neither indicated what the specific effects were, nor stated whether other toxic effects were seen at the same or lower dose levels, nor explained how the levels used in these studies compared with real-world exposure levels. The list was based largely on secondary references. Many of the agents appeared to be on the list only because they have been shown to have estrogenic activity in one or more in vitro (test tube) assay systems. The concern was that the listed chemicals, by binding with the estrogen receptor, may act like the synthetic estrogen, diethylstilbestrol. Despite the deficiencies in the research supporting the list, it has been republished and expanded in many subsequent publications. It has even been used by regulatory agencies, including a 1997 publication by the state of Illinois that categorizes chemicals into "known, probable, and suspect" endocrine disruptors. The lay press has presented concerns about endocrine disruption in the print and electronic media. Assault on the Male is a documentary made for the BBC television show Horizon in 1993. The documentary was written by Deborah Cadbury, who later wrote a book entitled The Feminization of Nature *(3)*. The documentary presents scientists, such as Dr. John McLachlan, Dr. Louis Guillette, Dr. Richard Sharpe, Dr. Ana Soto, and Dr. Theo Colborn, to explain their concerns about hormonally active chemicals. The story is both alarming and compelling. It was aired in Europe first and then in the United States, and helped stimulate interest in the potential effects on human0s and wildlife.

Dr. Colborn's concerns about the environmental endocrine issue, as well as the basis for those concerns, are presented in a book entitled Our Stolen Future (1996) *(4)*. The presentation relies heavily upon researchers that concur with Dr. Colborn's views on this issue, and this book has captured the attention of many politicians and the media. The foreword was written by Vice President Al Gore, and he acclaims the book as being as significant as Rachel Carson's Silent Spring.

Although Our Stolen Future has increased the public's and the media's attention to this issue, political action has been part of the environmental endocrine issue since 1994. At that time, the Clinton Administration's Clean Water Initiative *(5)* proposed legislation that would have led to the study and phaseout of chlorine because of a concern that persistent chlorinated chemicals could adversely affect human health, particularly the endocrine system. Although that bill did not pass, other laws have been enacted that address the environmental endocrine issue. The

Food Quality Protection Act of 1996 (FQPA) and the Safe Drinking Water Act of 1996 (SDWA) each include provisions requiring the screening of chemicals for estrogenic activity and allowing the study of other hormonal activity for pesticides, inert ingredients in pesticides, and other environmental chemicals.

The passage of this legislation gave rise to the Endocrine Disruptor Screening and Testing Committee (EDSTAC), an advisory group that is helping the Environmental Protection Agency (EPA) determine how to screen and test chemicals for hormonal and anti-hormonal activity. The FQPA and the SDWA allow EPA to screen pesticides and other chemicals for estrogenic and other hormonal activity. EDSTAC is advising EPA on what hormonal activities should be investigated, and how this should be accomplished. At this time, EPA intends to consider screening many of the nearly 87,000 chemicals used in commerce for estrogenic, androgenic, and thyroid hormone agonist and antagonist activity. EDSTAC is proposing a system for setting priorities for screening and testing under which very rapid assay methods would be used for the first step of screening. EPA is developing the screening and testing system under a two-year regulatory deadline that began when FQPA was passed in August 1996.

Special Areas of Interest

"Endocrine disruption" is a broad term that generally implies that adverse effects are caused to an intact organism through a chemical effect on hormone synthesis, release, transport, metabolism, uptake, or action. The precise mechanism of action need not be known under most definitions. The "European Workshop on the Impact of Endocrine Disrupters on Human Health and Wildlife" defined an endocrine disruptor as "an exogenous substance that causes adverse health effects in an intact organism, or its progeny, secondary to changes in endocrine function" *(6)*. Initially, the EPA EDSTAC had considerable disagreement on the definition of an "endocrine disruptor." As a consensus definition, EDSTAC concluded that an endocrine disruptor is "an exogenous chemical substance or mixture that alters the structure or function(s) of the endocrine system and causes adverse effects at the level of the organism, its progeny, populations, or subpopulations of organisms, based on scientific principles, data, weight-of-evidence, and the precautionary principle." Many potential mechanisms and targets exist under these definitions. This paper identifies some of the most prominent or well recognized potential modes of action. Although it is not a complete list, the selection should provide the reader with an appreciation for the scope of issues faced in this area.

Hormone Synthesis, Release, and Transport, as Well as Hormone-Receptor Interaction. The endocrine system is one of the most complex systems in the human body. It controls and coordinates many basic functions as the body grows from a fetus through mature adulthood to old age. The endocrine system provides homeostatic control for various systems. It also influences and controls the development of organ systems. The endocrine system includes the brain, reproductive organs, and other glands; the hormones they secrete; and the receptors in target organs that respond to the hormones. Through these chemical messengers, the endocrine system communicates with the body, including the reproductive, immune, nervous, respiratory, and digestive systems.

The interest in the environmental endocrine issue has focused on three major hormone systems: The ability of chemicals to act as estrogen (E), androgen (A), or thyroid (T) agonists or as antagonists has been the primary focus in this issue. The EPA EDSTAC is defining a screening battery that will identify chemicals that interact with these three hormone receptor systems. Each hormone system is controlled by the pituitary gland. Estrogens (estradiol, estriol, and estrone), progestins (progesterone), and androgens (testosterone and dihydrotestosterone) are gonadal steroid hormones. Their synthesis and release are controlled by pituitary peptide hormones: follicle stimulating hormone (FSH) and luteinizing hormone (LH). FSH and LH are under control, in turn, of hypothalamic gonadotrophin releasing hormone (GnRH). GnRH, LH, and FSH synthesis and release are all affected by steroid hormone levels. This hypothalamic-pituitary-gonadal axis is a feedback loop that maintains appropriate hormone levels. This system supports gametogenesis, accessory sex organ function, and reproductive system development. Depending upon the type and timing of an effect, disruption could temporarily or permanently alter these organ systems.

In Utero Exposure and Reproductive Tract Function and Development. The greatest concerns for endocrine effects are those that cause permanent changes during development that may not be detected until the affected organism reaches sexual maturity. DES has been proposed as the model for those concerns, although the potency and dose of DES is generally much greater than that for other chemicals being postulated as potential environmental endocrine disruptors. DES is an example of an agent that has caused significant effects through a hormonal mechanism. The consequences of that exposure were profound .

DES was administered to approximately four million pregnant women, between 1938 and 1971, to prevent miscarriages. As adults, the children of the women who took DES had increased rates of reproductive cancers and other gonadal diseases. The DES-exposed daughters have increased incidences of clear cell adenocarcinoma, vaginal adenosis, infertility, and other vaginal epithelial changes. While DES-exposed sons have not experienced increased rates of cancer they have experienced anomalies of the genital tract.

Breast Cancer. The American Cancer Society estimates 180,200 new breast cancer cases in women in the United States in 1997. Since 1987 (based on data from 1993), the age-adjusted breast cancer rates have leveled off at about 110 cases per 100,000 women, after increasing steadily from 83 per 100,000 in 1973 (NCI Surveillance, Epidemiology, and End Results Data *(7)*. The increase in the number of breast cancer cases of about 4 percent per year has been attributed largely, but not completely, to improved diagnosis. Mortality has remained between 25 and 28 per 100,000 women since 1973 and has not changed significantly, despite improved detection.

Estrogens may have a paradoxical effect on breast cancer. On the one hand, some risk factors for increased breast cancer appear to be related to increased endogenous estrogen exposure. On the other hand, weak exogenous estrogens, like phytoestrogens produced by plants such as soy beans, appear to be associated with decreased breast cancer risk *(8-9)*. The incidence of breast cancer in Japan is much lower than that in the United States. Many doctors believe the large amounts of soy protein in the typical Japanese diet significantly lower cancer risk. Paradoxically, soy protein is rich in naturally occurring estrogenic substances. More research on the mechanism of the body's response to naturally occurring estrogenic substances from plants and other sources is needed to understand the differences in cancer rates between Japan and the United States, including ethnic differences.

Factors are present in a soy-rich diet that decrease breast cancer risk *(10-11)*. Soy contains isoflavone aglycones, like daidzein, equol, and genistein, which have estrogenic activity. They also have anti-estrogenic activity and other potentially anti-carcinogenic activity. Soy has been shown to have protective effects against breast cancer, as well as other types of cancer. Are chemicals in soy acting as weak estrogens or anti-estrogens that limit the potential adverse effects of more potent estrogens? Or are they acting as anti-carcinogens by inhibiting protein tyrosine kinase or other critical enzymes involved in signal transduction? These are important questions that may affect drug design and dietary advice related to phytoestrogens.

Male Reproductive Effects: Sperm Count, Fertility, and Accessory Sex Organs. Of all the recent controversies about the potential effects of estrogenic chemicals on the human body, the apparent decline in sperm count has generated the most attention—and the most confusion *(12-15)*. Several studies in the past few years have reported that, in many countries of the world, levels of sperm produced by adult males are declining.

One study published in 1992 in the British Medical Journal concluded that the worldwide level of sperm production had declined 50 percent during the past 50 years *(16)*. That study was a statistical analysis combining data from 61 different studies on male semen quality (sperm density and semen volume). The 61 reports were published between 1938 and 1990. The meta-analysis of those studies showed a highly significant ($p<0.0001$) decline in sperm count from 133×10^6 sperm/mL in 1940 to 66×10^6 sperm/mL in 1990 by linear regression analysis. Semen volume was also significantly lower, dropping from 3.40 to 2.75 mL from 1940 to 1990 ($p=0.027$) *(17)*. The men in the 61 studies were semen donors either of proven fertility in some studies, or of unknown fertility in others. Studies were from various countries; 28 studies were from the United States. The group sizes varied from 7 to 4435 men per study. The authors correlate the drop in semen quality with increases in testicular cancer, cryptorchidism, and hypospadias. They have developed a hypothesis that these changes could be related to increased environmental contaminants, especially xenoestrogens *(17)*. The hypothesis that was repeated by others is that estrogens could be responsible for a drop in sperm count and a decline in male reproductive health *(18, 14)*. Jensen et al. described a decline in male reproduction since World War II, and speculated that the decline was related to the use of DES by pregnant women, or exposure to environmental estrogens. Sharpe and Skakkebaek *(14)* also hypothesized that the male effects could be the result of estrogen exposure. Changes in male reproductive tract development have been observed in humans and animals exposed to DES. Prenatal exposure to estrogens can alter mullerian duct regression, leading to male reproductive tract abnormalities. Leydig cell and Sertoli cell activity can also be affected, which could change testosterone production and sperm production, respectively.

The Carlsen study raised important issues by looking at old data in a new way. It was argued, however, that there were other ways to look at the same data. When the data from the 61 studies were reanalyzed by other scientists, a different conclusion was reached *(19)*. The original analysis used a linear model to test for a trend in sperm counts. Olsen and coworkers used three other statistical models: the quadratic, spline fit, and stairstep. They provided six reasons that support their view that the linear model was not the best choice. They concluded that, if the initial decline was real, there was no further decline after 1960. In fact, the most robust data are found in the last 20 years, and they show constant or slightly increasing counts.

Explanations of the complexities of measuring sperm count, and the potential influence of the factors on the data, are provided by Lerchl and Nieschlag *(12)*. They point out that such studies were based on evaluating sperm counts in normal men. Normal, however, is defined by the World Health Organization (WHO) with a numerical cut-off, which has changed (decreased) over the years. That drop in the line separating normal from abnormal could account for a drop in "normal" sperm count, if the laboratory censures all subnormal values from the database, or rejects subnormal men as sperm donors *(12)*. Bromwich et al. *(20)* studied the influence of changing the reference value for normal, and investigated whether sperm count distribution was better modeled by a normal distribution or a lognormal distribution. They concluded that the reported decline in sperm concentration could be an artifact of the methods used by Carlsen et al. (12) and also identified important issues that have to be controlled, such as age of donor, duration of abstinence prior to donation, and comparability of methods among laboratories.

The Carlsen et al. study concluded that there was a global reduction in sperm count. Other investigators have narrowed their focus to specific locations. One of the strongest studies was by Auger et al. *(21)*. They reported a decline in sperm count in Paris, France, between 1973 and 1992. They studied 1351 fertile men, and mean sperm concentration dropped approximately 2.1 percent per year, from 89×10^6 sperm/mL to 60×10^6 sperm/mL ($p<0.001$). Studies in Toulouse, France, from 1977 to 1992 failed to observe a decline like that seen in Paris *(22)*.

A study of 1283 U.S. men showed no decline in semen quality in various regions of the United States from 1970 to 1994 *(23)*. There was a slight increase in mean sperm concentration, but not in sperm motility or semen volume. There were significant regional differences: New York had the highest sperm concentrations and motility, Minnesota had the second-highest, and California had the lowest. No explanation for the geographic differences has been provided, but different counting chambers were used in California than in New York and Minnesota. A seven-year study from 1979 to 1986 showed no decline in daily sperm production samples from Texas *(24)*. Data from Seattle, Washington, showed no decline in semen quality from 1972 to 1993, using 510 healthy men *(25)*.

It has been hypothesized that declines in testicular function or sperm count could be related to estrogen exposure *(14)*. However, few, if any, sperm count studies have included data on the extent to which participants were exposed to chemicals in the environment. Thus, the research provides no basis for suggesting a link between chemical exposure and sperm counts, and the link to estrogens is highly tenuous, but it is biologically plausible. Scientists do not understand the factors that cause sperm counts to differ greatly from city to city or among neighboring countries. At this time, the pattern of variation does not seem to be consistent with the idea that exposure to chemicals in the environment—either before or after birth—is affecting sperm count.

It has been suggested that in utero exposure to synthetic chemicals with estrogen-like properties could be a risk factor for testicular cancer *(14, 26-27)*. Some studies of men whose mothers took DES during pregnancy have found a possible increased risk of testicular cancer and of undescended testes at birth *(28-29)*.

Prostate Cancer. Accessory sex organs, like the prostate, are generally derived from the Wolffian duct. Part of the prostate, called the prostatic utricle, is a remnant of the Mullerian duct. That remnant can persist with prenatal exposure to DES, and may give rise to prostate disease *(28)*. Cancer of the prostate gland is the second most prevalent form of cancer. The most significant risk factor for prostate cancer is age, and other established risk factors are related to family history of the disease, ethnicity, and country of residence *(30)*. The prostate is a hormone-responsive tissue. Exogenous estrogens could be causing higher rates of cancer of the prostate gland by interfering with the endocrine system.

Like breast cancer, prostate cancer rates increase dramatically with age. The SEER data showed 43 cases of prostate cancer per 100,000 for men under 65 and 1239 per 100,000 for men over 65 from 1990 to 1994 *(7)*. Also, like breast cancer, the incidence of prostate cancer has shown a dramatically increasing trend with a 294 percent increase in men under 65 from 1973 to 1994, and a 110 percent increase in prostate cancer in men over 65 *(7)*. After increasing slowly for many years, the incidence of new cases jumped in 1987 and has grown approximately 20 percent each year since then. However, this increase is primarily the result of better detection of the disease through increased screening. By using new lab tests and ultrasound techniques, early-stage prostate cancers are detected that could not

be diagnosed before. In other words, the risk of getting prostate cancer may not have changed, but the chances of a doctor discovering a cancer early enough to treat it have improved *(31)*.
The growth of prostate cancers can often be controlled by reducing the level of testosterone in the blood, suggesting the cancer is sensitive to the hormone. However, research has consistently failed to show any link between levels of hormones in the bloodstream and risk of the disease. The potential connection between prostate cancer and exposure to estrogenic chemicals in the environment is an area that needs further study.

Interesting experimental findings on prostate weight changes, but not on cancer or disease, have shown that very low doses of DES in utero will increase prostate weight slightly *(32)*. Higher doses of DES decreased prostate weight. The experiment was conducted on a very small sample size, and the weight increases were also small. No histopathology has been presented to describ the morphological changes, but these studies have been presented as a model for estrogeninduced prostate cancer and disease.

Screening, Testing, and Risk Assessment Issues. The EDSTAC Final Report *(33)* spells out a series of short-term mechanistic studies that are included under the high throughput pre-screen (HTPS) and Tier 1 Screen (T1S). The report also describes the studies that could be useful for safety assessment under Tier 2 Testing (T2T).

Risk assessment is based upon hazard, potency, and exposure. Hazard is the inherent property of a chemical to cause an adverse effect. For chemicals that affect the endocrine system, it is not a simple task to identify such adverse effects. Because these chemicals act on systems that affect normal homeostatic responses, it is not always clear whether a change in an end point will necessarily lead to a change that adversely affects structure or function. Many of the assay systems that are currently being proposed to screen for endocrine activity are the same screens that would identify physiological or pharmacological activity. Estrogen and androgen hormone receptor binding have both been proposed as part of the screen for potential endocrine-disrupting activity. Effects on hormone receptor binding and hormone synthesis have both been proposed to be included in a screening battery. The screening assessment is based on physiological activity and the ability of a chemical to mimic or to antagonize hormone activity. Adverse effects would be considered in later tests. Compounds should not be labeled as endocrine disruptors based on the results of screening tests.

Testing methods have been reviewed in a number of articles *(34-36)*. Methods have been developed to assess broad biological functions, as well as specific molecular events, related to endocrine function and toxicity. Test selection is critical.

The potency of hormonally active agents can differ by orders of magnitude *(37)*. Estradiol and DES are potent estrogens, while plant estrogens may be one to several orders of magnitude less potent [See Table 2] *(38)*. Synthetic chemicals that have received so much attention are often even less potent by several orders of magnitude.

For example, considerable attention has been focused on pesticides, because DDT was one of the first substances to be identified as hormonally active. However, DDT and several other hormonally active pesticides have very weak estrogenic potency.

Moreover, pesticides introduced in recent years all have been tested to determine their effect on reproduction. Scientists specifically look for evidence of reproductive effects during extensive testing of pesticide active ingredients before they are registered. Any endocrine-related effects would probably be detected during this testing.

Among the thousands of other chemicals used every day, relatively few have been tested as thoroughly as pesticides *(39)*. Industry, government, and environmental groups are working together to come up with a practical and scientifically valid way to screen chemicals for endocrine disruption potential and conduct follow-up testing if needed. Currently there are no generally accepted, validated methods to screen chemicals for possible hormonal activity that might lead to adverse health effects because the endocrine system is so complex. However, rapid progress is being made.

In cases where hormonal activity is detected in a simple screening test there is no proof that a chemical is harmful; such a finding indicates the need for further testing using more complex methods *(35)*.

Finally, the risk to humans or to wildlife involves comparing the hazard and potency to the levels actually found in the environment. This type of evaluation has been heavily influenced by lists of endocrine disrupting chemicals that do not consider exposure or potency *(2, 40)*. However, complete safety assessment and regulatory action must consider exposure to hormonally active agents, which must be studied in various environmental media to fully characterize the risks of these agents.

Dose-Response Models, Synergism, and Threshold. An area of hot debate is how the dose response of hormonally active agents should be modeled *(41-42)*. The questions are as follows: Is there a threshold for hormone-receptor mediated responses? Is the dose response curve fundamentally different than for other toxicants? Are responses common at levels below the conventionally determined no-observed-adverse-effectlevel (NOAEL)? How do plasma binding proteins affect the dose response curve?

Major controversy exists regarding whether the addition of a hormonally active chemical will automatically add to an endogenous response. To date, regulators have treated hormonally active agents as having to reach a threshold response before a biological or adverse effect would be observed, and they appear to be holding to that position at this time *(43)*. However, other investigators hold the view that, because the endocrine system is already "turned on," any additional hormone will increase the background response *(42)*. However, it has been argued that such a view is only supported if the ligand is the natural hormone because the uptake, metabolism, receptor binding, and antagonism are different for each ligand *(41)*.

It has also been reported that high-dose toxicology studies have the potential to miss effects because the dose selection may overshoot low level effects. Changes in prostate weight were observed at low prenatal doses of DES and bisphenol A, which disappeared at higher levels, as the prostate first increased, then decreased in weight *(44)*. Those studies, however, were conducted on small group sizes (n = 6), and histopathology was not reported, which could have explained why the organ weight increased and then decreased. The research is currently being repeated by other investigators.

Studies on synergism that were published with considerable fanfare *(45)* demonstrate the importance that such critical work be verified and repeated. The original publication described a combination of pesticides and reported a 1,600-fold greater-than-expected response for hormone binding and in vitro response in a yeast cell assay. The work was especially surprising because of the magnitude of the reported synergism, and the finding that two similar chemicals could cause such an extraordinary response. It is more typical to see a synergistic response when a target is hit from two different modes of action that enhance each other. However, subsequent attempts by other laboratories to repeat the work failed *(46-47)*. Ultimately, the original authors acknowledged that the work was flawed and withdrew their paper *(48)*. At this point, it is generally assumed that multiple chemical interactions will either be antagonistic, or additive, but they are unlikely to be significantly more than additive.

Conclusion. A great deal of media and scientific attention is being paid to the issue of endocrine disruption. Significant scientific issues continue to be researched, while the U.S. EPA is developing a screening system, as required by the U.S. Congress. The testing program that has been proposed by the EDSTAC will undergo a validation program that may lead to significant changes in the design of the testing program.

As currently designed, the program will first evaluate whether or not the chemicals have inherent hormonal activity. Priorities will then be set for further testing and risk assessment. The T2T testing designs will include endpoints that are sensitive to endocrine and nonendocrine mechanisms of action. Therefore, adverse effects that are unrelated to endocrine disruption may be used for setting safe levels.
It is important to note that positive findings may not be appropriate to label a chemical as an "endocrine disruptor" because the adverse effects may be from another mechanism. However, such an approach should ultimately lead to greater confidence in the safety assessment process.

Literature Cited

1) Colborn, T.; Clement, C. "Chemically-Induced Alterations in Sexual and Functional Development: The Wildlife/Human Connection." *Advances in Modern Environmental Toxicology*; Editor, M. A. Mehlman; Princeton Scientific Publishing Co., Inc.: Princeton, New Jersey, **1992**; p. 403.
2) Colborn, T.; vom Saal, F. S.; Soto, A. M. "Developmental Effects of Endocrine-Disrupting Chemicals in Wildlife and Humans." *Environmental Health Perspectives,* **1993**, *vol. 101(5),* pp. 378-384.
3) Cadbury, D. *The Feminization of Nature: Our Future at Risk;* Hamish Hamilton: London, **1997**.
4) Colborn, T.; Dumanski, D.; Myers, J.P. *Our Stolen Future: Are We Threatening Our Fertility, Intelligence, and Survival? A Scientific Detective Story.* New York, Dutton, Penguin, USA, 1996.
5) Office of Water. President Clinton's Clean Water Initiative: EPA8000R94001; U.S. Environmental Protection Agency: Washington, D.C., 1994.
6) MRC Institute for Environment and Health. *Report of Proceedings*. European Workshop on the Impact of Endocrine Disrupters on Human Health and Wildlife; Weybridge, UK, **1996**, pp. 1-109.
7) Ries, L. A. G.; Kosary, C. L.; Hankey, B. F.; Miller, B. A.; Harras, A.; Edwards, B. K. (eds.). *SEER Cancer Statistics Review, 1973-1994*. NIH Pub. No. 97-2789; National Cancer Institute: Bethesda, Maryland, 1997.
8) Adlercreutz, H. "Phytoestrogens: Epidemiology and a Possible Role in Cancer Protection." *Environmental Health Perspectives*, **1995**, *vol. 103*, pp. 103-112.
9) Adlercreutz, H.; Honjo, H.; Higashi, A.; Fotsis, T.; Hamalainen, E.; Hasegawa, T.; Okada, H. "Urinary Excretion of Lignans and Isofavonoid Phytoestrogens in Japanese Men and Women Consuming a Traditional Japanese Diet." *American Journal of Clinical Nutrition*, **1991**, *vol. 54*, pp. 1093-1100.
10) Barnes, S.; Grubbs, C.; Setchell, K. D. R.; Carlson, J. "Soybeans Inhibit Mammary Tumors in Models of Breast Cancer." *Mutagens and Carcinogens in the Diet*, **1990**, pp. 239-253.
11) Messina, M. J.; Persky, V.; Setchell, K. D. R.; Barnes, S. "Soy Intake and Cancer Risk: A Review of the In Vitro and In Vivo Data." *Nutrition and Cancer*, **1994**, *vol. 21(2)*, pp. 113-131.
12) Lerchl, A.; Nieschlag, E. "Decreasing Sperm Counts? A Critical (Re)view." *Experimental and Clinical Endocrinology and Diabetes*, **1996**, *vol. 104,* pp. 301-307.
13) Lemonick, M. D. "What's Wrong with Our Sperm?" *TIME Magazine*, **1996**, *vol. 147*, p. 12.
14) Sharpe, R. M.; Skakkebaek, N. E. "Are Oestrogens Involved in Falling Sperm Counts and Disorders of the Male Reproductive Tract?" *The Lancet*, **1993**, *vol. 341*, pp. 1392-1395.
15) Swan, S. H.; Elkin, E. P.; Fenster, L. "Have Sperm Densities Declined? A Reanalysis of Global Trend Data." *Environmental Health Perspectives*, **1997**, *vol. 105(11),* pp. 1228-1232.

16) Carlsen, E.; Giwercman, A.; Keiding, N.; Skakkebaek, N. E. "Evidence for Decreasing Quality of Semen During Past 50 Years." *BMJ*, **1992**, *vol. 305*, pp. 609-613.
17) Carlsen, E.; Giwercman, A.; Keiding, N.; Skakkebaek, N. E. "Declining Semen Quality and Increasing Incidence of Testicular Cancer: Is There a Common Cause?" *Environmental Health Perspectives*, **1995**, *vol. 103 (Suppl. 7),* pp. 137-139.
18) Jensen, T. K.; Toppari, J.; Keiding, M. N.; Skakkebaek, N. E. "Do Environmental Estrogens Contribute to the Decline in Male Reproductive Health?" *Clinical Chemistry*, **1995**, *vol. 41(12),* pp. 1896-1901.
19) Olsen, G. W.; Bodner, K. M.; Ramlow, J. M.; Ross, C. E.; Lipshultz, L. I. "Have Sperm Counts Been Reduced 50 Percent in 50 Years? A Statistical Model Revisited." *Fertility and Sterility*, **1995**, *vol. 63(4),* pp. 887-893.
20) Bromwich, P.; Cohen, J.; Stewart, I.; Walker, A. "Decline in Sperm Counts: An Artefact of Changed Reference Range of 'Normal'?" *BMJ*, **1994**, *vol. 309*, pp. 19-22.
21) Auger, J.; Kunstmann, J. M.; Czyglik, F.; Jouannet, P. "Decline in Semen Quality Among Fertile Men in Paris During the Past 20 Years." *The New England Journal of Medicine*, **1995**, *vol. 332(5),* pp. 281-285.
22) Bujan, L.; Mansat, A.; Pontonnier, F.; Mieusset, R. "Time Series Analysis of Sperm Concentration in Fertile Men in Toulouse, France Between 1977 and 1992." *BMJ*, **1996**, *vol. 312*, pp. 471-472.
23) Fisch, H.; Goluboff, E. T.; Olson, J. H.; Feldshuh, J.; Broder, S. J.; Barad, D. H. "Semen Analyses in 1,283 Men from the United States over a 25-Year Period: No Decline in Quality." *Fertility and Sterility*, **1996**, *vol. 65(5),* pp. 1009-1014.
24) Johnson, L.; Levine, R.; Barnard, J. J.; Neaves, W. B. "No Decline in Daily Sperm Production in a Group of North American Men Over a Seven-Year Period." *Journal of Andrology*, **1996**, *vol. 78*, p. 42.
25) Paulsen, C. A.; Berman, N. G.; Wang, C. "Data from Men in Greater Seattle Area Reveals No Downward Trend in Semen Quality: Further Evidence that Deterioration of Semen Quality is not Geographically Uniform." *Fertility and Sterility*, **1996**, *vol. 65(5*), pp. 1015-1020.
26) Toppari, J.; Larsen, J. C.; Christiansen, P.; Giwercman, A.; Grandjean, P.; Guillette Jr., L. J.; Jegou, B.; Jensen, T. K.; Jouannet, P.; Keiding, N.; Leffers, H.; McLachlan, J. A.; Meyer, O.; Muller, J.; Meyts, E. R.-D.; Scheike, T.; Sharpe, R.; Sumpter, J.; Skakkebaek, N. E. *Male Reproductive Health and Environmental Chemicals with Estrogenic Effects,* Ministry of Environment and Energy, Danish Environmental Protection Agency: Copenhagen, 1995.
27) Toppari, J.; Larsen, J. C.; Christiansen, P.; Giwercman, A.; Grandjean, P.; Guillette Jr., L. J.; Jegou, B.; Jensen, T. K.; Jouannet, P.; Keiding, N.; Leffers, H.; McLachlan, J. A.; Meyer, O.; Muller, J.; Rajpert-De Meyts, E.; Scheike, T.; Sharpe, R.; Sumpter, J.; Skakkebaek, N. E. "Male Reproductive Health and Environmental Xenoestrogens." *Environmental Health Perspectives*, **1996**, *vol. 104(Suppl. 4),* pp. 741-803.
28) Newbold, R. "Cellular and Molecular Effects of Developmental Exposure to Diethylstilbestrol: Implications for Other Environmental Estrogens." *Environmental Health Perspectives*, **1995**, *vol. 103(Suppl. 7),* pp. 83-87.

29) U.S. Department of Health and Human Services. *NIH Workshop Long-Term Effects of Exposure to Diethylstilbestrol (DES).* NIH Workshop Long-Term Effects of Exposure to Diethylstilbestrol (DES), Publishing Agency: Falls Church, Virginia, 1992, pp. 1-94.
30) Key, T. "Risk Factors for Prostate Cancer." *Cancer Surveys*, **1995**, *vol. 23*, pp. 63-77.
31) Lee, W. R.; Giantonio, B. J.; Hanks, G. E. "Prostate Cancer." *Current Problems in Cancer*, **1994**, *vol. 18(6)*, pp. 295-357.
32) vom Saal, F. S.; Timms, B. G.; Montano, M. M.; Palanza, P.; Thayer, K. A.; Nagel, S. C.; Dhar, M. D.; Ganjam, V. K.; Parmigiani, S.; Welshons, W. V. *Prostate Enlargement in Mice Due to Fetal Exposure to Low Doses of Estradiol or Diethylstilbestrol and Opposite Effects at High Doses.* Proc. Nat'l. Acad. Sci.,, **1997**, *vol. 94*, pp. 106.
33) Endocrine Disruptor Screening and Testing Advisory Committee. *Final Report: Volumes I and II*; U.S. Environmental Protection Agency: Washington, D.C., 1998.
34) Organization for Economic Cooperation and Development. Draft Detailed Review Paper: Appraisal of Test Methods for Sex Hormone Disrupting Chemicals. Paris, France, 1997.
35) Reel, J.; Lamb, J.; Neal, B. *Survey and Assessment of Mammalian Estrogen Biological Assays for Hazard Characterization.* Jellinek, Schwartz, & Connolly, Inc.: Arlington, Virginia, 1996.
36) Zacharewski, T. "In Vitro Bioassays for Assessing Estrogenic Substances." *Environmental Science & Technology*, 1997, *vol. 31(3)*, pp. 613-623.
37) Safe, S. H. "Environmental and Dietary Estrogens and Human Health: Is There a Problem?" *Environmental Health Perspectives*, **1995**, *vol. 103(4)*, pp. 346-351.
38) Daston, G. P.; Gooch, J. W.; Breslin, W. J.; Shuey, D. L.; Nikiforov, A. I.; Fico, T. A.; Gorsuch, J. W. "Environmental Estrogens and Reproductive Health: A Discussion of the Human and Environmental Data." *Reproductive Toxicology*, **1997**, *vol. 11(4)*, pp. 465-481.
39) Stevens, J. T.; Tobia, A.; Lamb, J.; Tellone, C.; and O'Neal, F. *FIFRA Subdivision F Testing Guidelines: Are These Tests Adequate to Detect Potential Hormonal Activity for Crop Protection Chemicals?*, 1996.
40) Illinois Environmental Protection Agency. *Endocrine Disruptors Strategy.* Illinois EPA, 1997.
41) Lamb, J. C. "Can Today's Risk Assessment Paradigms Deal with Endocrine Active Chemicals?" *Risk Policy Report*, **1997**, *vol. 30*, pp. 32-35.
42) Sheehan, D. M.; vom Saal, F. S. "Low Dose Effects of Endocrine Disruptors - A Challenge for Risk Assessment." *Risk Policy Report*, **1997**, *vol. 31*, pp. 35-39.
43) Crisp, T. M; Clegg, E. D.; Cooper, R. L. *Special Report on Environmental Endocrine Disruption: An Effects Assessment and Analysis.* EPA/630/R-96/012; U.S. Environmental Protection Agency: Washington, D.C., 1997.
44) vom Saal, F. S.; Timms, B. G.; Montano, M. M.; Palanza, P.; Thayer, K. A.; Nagel, S. C.; Dhar, M. D.; Ganjam, V. K.; Parmigiani, S.; Welshons, W. V. "Prostate Enlargement in Mice Due to Fetal Exposure to Low Doses of Estradiol or Diethylstilbestrol and Opposite Effects at High Doses." Div. of Biological Sciences, University of Missouri: Columbia, Missouri, 1997.

45) Arnold, S. F.; Klotz, D. M.; Collins, B. M.; Vonier, P. M.; Guillette Jr., L. J.; McLachlan, J. A. "Synergistic Activation of Estrogen Receptor with Combinations of Environmental Chemicals." *Science*, **1996**, *vol. 272*, pp. 1489-1492.
46) Ramamoorthy, K.; Wang, F.; Chen, I.-C.; Norris, J. D.; McDonnell, D. P.; Gaido, K. W.; Bocchinfuso, W. P.; Korach, K. S.; and Safe, S. *Estrogenic Activity of a Dieldrin-Toxaphene Mixture in the Mouse Uterus, MCF-7 Human Breast Cancer Cells and Yeast-Based Estrogen Receptor Assays: No Apparent Synergism*, 1996.
47) Ramamoorthy, K.; Wang, F.; Chen, I.-C.; Safe, S.; Norris, J. D.; McDonnell, D. P. Gaido, K. W.; Bocchinfuso, W. P.; Korach, K. S. "Potency of Combined Estrogenic Pesticides." *Science*, **1997**, *vol. 275*, pp. 405-406.
48) McLachlan, J. A. "Synergistic Effect of Environmental Estrogens: Report Withdrawn." *Science*, **1997**, *vol. 277(5325),* pp. 462-463.

Chapter 5

Aggregate and Cumulative Exposure and Risk Assessment

Charles B. Breckenridge [1], Robert L. Sielken, Jr. [2], and James T. Stevens [1]

[1] Novartis Crop Protection, Inc., P. O. Box 18300, Greensboro, NC 27419-8300
[2] Sielken, Inc., 3833 Texas Avenue, Bryan, TX 77802

Interest in methodology for assessing the probability of exposure to a single chemical arising from multiple pathways (e.g. diet, water, residential) or to multiple chemicals having the same mechanism of toxicity has increased since the Food Quality Protection Act became law in 1996. Use of probability distributions to characterize exposure make it possible for 1) the continuum of data from the largest to the smallest values to be expressed, 2) the relative likelihood of occurrence to be described, 3) the uncertainty for each component to be reflected and for 4) the individual variability in the population to be captured. Exposure can be aggregated in a mathematically correct way and be characterized relative to benchmarks of toxicity such as the NOEL, the RfD, the ED_{10} or an upper bound cancer potency estimate (Q_1*). Using these procedures, the risk manager can determine the probability that exposure is less than or equal to an acceptable daily dose for the whole population or a selected subpopulation.

The Food Quality Protection Act (1) now mandates that the US Environmental Protection Agency consider the aggregate exposure and the associated risk of single chemical exposure arising from multiple sources (i.e. diet, water and non-occupational sources) and route (i.e. oral, dermal, inhalation). Furthermore, when two or more chemicals share a common mechanism of toxicity, then the cumulative dose from exposure to these chemicals must be estimated. In this paper a three-tier assessment approach is proposed. Tier 1 uses default assumptions and single point (deterministic) estimates of exposure, hazard and risk employing procedures routinely used by EPA and the crop protection industry (2). In Tier 2, combinations of deterministic and probabilistic (distributional) data are used, while Tier 3 assessments rely predominantly on distributional data.

Significant routes of exposure

In conducting an aggregate risk assessment for pesticides, three primary routes of exposure should be considered: oral ingestion, dermal absorption and inhalation. Ingestion includes dietary intake as well as the consumption of drinking water. Dermal exposure is primarily limited to dermal contact following the agronomic and residential use of pesticides, while inhalation exposure includes the breathing of volatile and nonvolatile constituents (dust) of pesticides either during or after residential use.

Exposure from Ingestion

Procedures for estimating exposure to pesticides from dietary sources have been developed by the EPA (3) (Dietary Residue Exposure System; DRES) and have provided the basis for past tolerance-setting decisions. Chronic dietary exposure to a pesticide is calculated by assuming that pesticide residues on food exist at tolerance levels, or at more realistic anticipated residue levels. Refinements to the analysis may take into account information on market share, food processing factors and studies that define the transfer of residues in fed commodities to milk, meat and eggs.

New guidance has been developed by EPA to assess the magnitude of pesticide exposure from food ingested during a single day (4). The single highest residue value, the average residue, the 95th percentile of a residue distribution, or the entire distribution of residues are used to derive a distribution of pesticide exposure for a sample population identified in the USDA continuing food intake survey for individuals (5).

An assessment of pesticide exposure from drinking water is less well developed. According to guidance established in the Primary Drinking Water Standard (6), EPA has set MCL's (maximum contaminant levels) or MCLG's (maximum contaminant level goals) for selected pesticides. Traditionally, 20%

of the acceptable daily intake of a pesticide is allocated to drinking water based on a daily water consumption of 2 liters for adults. Under the Food Quality Protection Act this general rule of thumb has been reconsidered such that Tier 1 analyses rely on the predicted concentration of pesticide residues in groundwater (7) or surface water (8). Refinements of these models have been proposed (9), and it has been suggested that higher tier analysis be based upon water monitoring data (10).

Dermal Exposure

The Food Quality Protection Act mandates that exposure from residential sources be combined with exposure from ingestion. It is expected that a fraction of residential exposure will result from dermal exposure secondary to indoor or outdoor residential pesticide application. The Outdoor Residential Exposure Task Force (11) has been organized to develop a database which will include exposure data for individuals applying pesticides to turf and subsequently re-entering the treated area as well as for bystanders that may enter the treated area at various time intervals post-application. A similar task force has been commissioned by industry to develop data that can be used to characterize pesticide exposure resulting from pesticide use in and around the home.

Inhalation Exposure

In most cases, the inhalation of pesticides is a minor route of exposure. Residential treatments are usually applied outside the home and the dilution in the atmosphere results in minimal opportunity for significant inhalation exposure. The exception to this is the use of pesticides in confined spaces such as termite treatment and fogging uses.

Decision Logic for Assessing Aggregate and Cumulative Exposure

Prior to the conduct of an aggregate risk assessment, it is recommended that the decision logic presented in Flow Chart 1 be used to determine potential exposure sources. Thus pesticide exposure from drinking water should be included in the Tier 1 screen if high solubility or mobility or low rates of environmental degradation suggest that the chemical may be found in drinking water sources. Likewise, an evaluation of dietary exposure should be considered if tolerances are required or residues on food are anticipated to be detected at quantifiable levels. Exposure from non-occupational exposure sources may be included in the Tier 1 screen if either outdoor or indoor residential uses are proposed.

An aggregate assessment for a single pesticide may be required if concomitant exposure from multiple sources is expected and a cumulative exposure assessment for two or more pesticides would be necessary if they meet the criteria of having a common mechanism of toxicity.

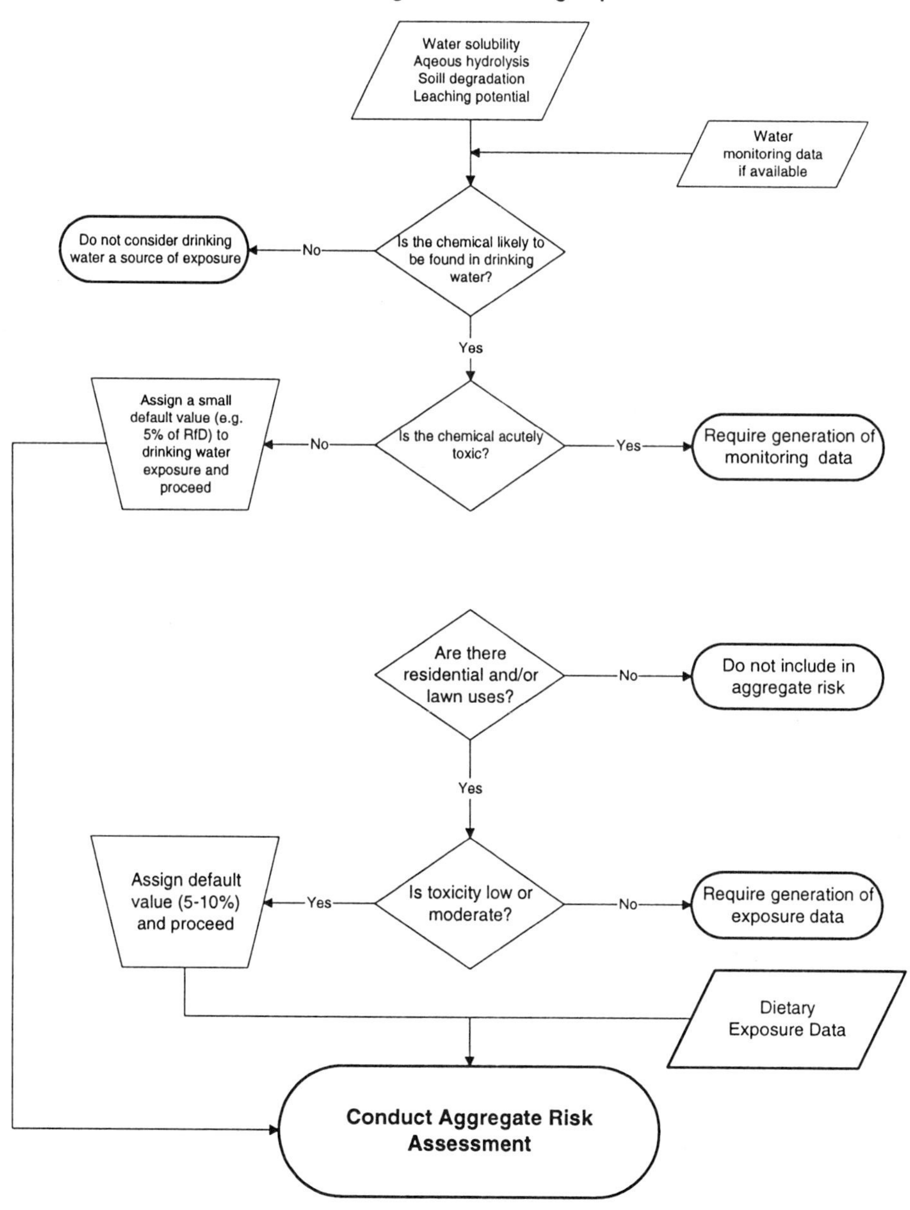
Decision Logic for Assessing Exposure
Water solubility
Aqeous hydrolysis
Soill degradation
Leaching potential
Water
monitoring data
if available
Is the chemical likely to be found in drinking water?
No
Do not consider drinking water a source of exposure
Yes
Is the chemical acutely toxic?
No
Assign a small default value (e.g. 5% of RfD) to drinking water exposure and proceed
Yes
Require generation of monitoring data
Are there residential and/or lawn uses?
No
Do not include in aggregate risk
Yes
Is toxicity low or moderate?
Yes
Assign default value (5-10%) and proceed
No
Require generation of exposure data
Dietary
Exposure Data
Conduct Aggregate Risk Assessment

Flow Chart 1.

AGGREGATE EXPOSURE/RISK ASSESSMENT

Definitions

Deterministic Estimate: Comprised of an estimate or upper bound estimate of the population mean (arithmetic or geometric), median, or percentile of exposure, hazard or risk.

Probability Distribution: Comprised of a distribution of values that define the magnitude of the exposure, hazard or risk. Each value is associated with a probability of occurrence.

Aggregate Exposure: Defined as the concurrent exposure to a single pesticide arising from multiple sources or routes. Aggregate exposure to a pesticide is calculated for the individual based upon exposure during a single day or the average daily exposure occurring over different durations of time including an estimate of the average lifetime daily dose. A distribution of such individual dose estimates can be constructed such that the probability that an individual might receive a specific dose can be determined.

Cumulative Exposure: Defined as the concurrent exposure to two or more pesticide residues or their metabolites. As with the aggregate exposure estimates, the cumulative exposure of an individual to multichemical pesticide residues may be based on the estimated dose occurring on a single day or on the average daily dose calculated over various durations of time up to and including the average lifetime daily dose. Cumulative exposure estimates can be represented either deterministically or probabilistically.

Acceptable Daily Dose: Defined as the mg/kg/day dose that is generally regarded as safe. The acceptable daily dose may be established for toxicity endpoints derived from acute, subchronic or chronic toxicity studies. Examples of acceptable daily doses that are either currently in use or proposed include the RfD (Reference Dose), and the VSD (Virtually Safe Dose or Risk Specific Dose (RSD); the dose that would lead to an added risk of 1 in a million).

Benchmark Dose: The benchmark dose is the dose that has no toxicological effect. The no observed effect level (NOEL), the no observed adverse effect level (NOAEL) and the ED_{10} (Dose that results in a 10% increase in the probablity of a response) and the LED_{10} (lower 95% confidence limits on the ED_{10}) are benchmark doses.

Uncertainty Factor: In a deterministic risk assessment, the benchmark dose derived from a suitable toxicity study is divided by the uncertainty factor to obtain the acceptable daily dose.

Percentile of the Risk Distribution: In a probabilistic risk assessment, a specific percentile of the risk distribution defines an acceptable level of risk.

Risk Allocation: Represents the proportion or percent of the reference dose that is taken up by a single exposure source in a multi-source risk analysis or by a single chemical in a multi-chemical risk analysis. For cancer risk analysis this may be expressed as added cancer risk.

Margin of Exposure: Defined as the ratio of the no observed effect level to exposure; this is called a margin of safety if an RfD is used in the numerator.

$$\text{Margin of Exposure (MOE)} = \frac{\text{Benchmark Dose (NOEL, ED}_{10}\text{)}}{\text{Exposure}}$$

Hazard Index: The hazard index is defined as the ratio of the reference dose (RfD) to exposure as follows:

$$\text{Hazard Index (HI)} = \frac{\text{Exposure}}{\text{RfD}}$$

Hazard indices greater than 1 are generally considered unacceptable. When HI = 1, the margin of exposure is the same magnitude as the uncertainty factor used to define the reference dose (RfD).

Toxicity Equivalency Factor:

For chemicals with a common mechanism of toxicity, it may be necessary to account for differences in relative potency in order to combine risks together in an appropriate manner. The EPA has used upper bound estimates of carcinogenic potency (Q_1*) for genotoxic carcinogens and Toxicity Equivalency Factors (TEF's) have been used for non-cancer endpoints as defined below:

$$TEF_A = NOEL_{RC} \div NOEL_A$$

The toxic equivalency of chemicals sharing a common mechanism is expressed relative to a reference chemical (RC) in the class. An ED_{10} could equally well be used to standardize the potency of the chemicals in the class relative to that of the selected reference chemical.

Rationale for Using a Tier Approach

A tier approach is recommended for evaluating aggregate exposure and risk for a single pesticide arising from multiple sources or for a class of pesticides that share a common mechanism of toxicity.

The primary reasons for recommending a tier approach are:

1. A screening level analysis (Tier 1) permit regulatory decision-making for the majority of cases while consuming relatively few resources.

2. Options are available for higher tier analysis in case Tier 1 fails. Tier 2 analyses use both deterministic and probabilistic methods while Tier 3 assessments rely primarily on probabilistic techniques.

3. In cases where sufficient hazard and exposure data already exist, higher tier assessments utilizing distributional analyses may be conducted to provide a more realistic assessment of exposure and risk in selected populations. Variability in the data and the uncertainty in inferences concerning exposure and risk to man are more apparent in such analyses because the probability of being exposed at specified doses can be estimated.

Multi-Tier Risk Assessment

Probabilistic methods require extensive hazard and exposure data that are often not readily available. In order to efficiently use scientific resources, the simpler Tier 1 screening method currently may be used for a preliminary assessment. By conducting sensitivity analyses, exposure, hazard and dose factors that are likely to make significant contributions to risk can be identified and research can be more effectively prioritized. Table 1 provides a list of factors commonly used in assessing the risk from exposure to pesticides arising from diet, water and non-occupational pathways. A more detailed listing of such factors and the values commonly assigned to them can be found in *The Exposure Factors Handbook* (12) developed and published by the EPA and *The Exposure Factors Sourcebook* (13) published by the American Industrial Health Council.

Tier 1 analyses of the risk due to exposure to a pesticide from diet, water and non-occupational sources can be calculated by using constants for the parameters listed in Table 1. This table does not list all parameters that might be utilized in a comprehensive risk assessment, only a representative few that are frequently encountered. Sensitivity analyses can be conducted using the Monte Carlo simulation method to determine the effect on the calculated risk when these parameters take on different values or distributional characteristics. Thus, the parameters that are thought to significantly impact risk can be identified and data can be collected for a higher tier analysis if the chemical fails to pass the Tier 1 screen.

Table 1 – Typical Exposure, Dose and Risk Assessment Factors

Exposure Factors	Tier 1 (Default Value)	Tier 2	Tier 3
Exposure Factors for Dietary & Water			
Residue Level	Diet: ($\sum$ Tolerance) Water: (MCL)	➠	Residue Distribution
Market Share	Constant (100%)		Variable (Year, Region)
Food Intake	Mean (DRES)		Distribution
Water Intake	Constant (2 liters/day)		Distribution
Exposure Duration	Continuous (Daily)		Pop. Linked Distribution
Population Linked	Constant (100%)		Link to Population
Exposure Factors for Residential & Lawn Uses			
Dislodgeable Residue o Turf o Residential	Constant (100%)	➠	Distribution
Penetration Factors o Clothing/Type o Dermal	 Constant (80%) Constant (100%)		 Distribution Experimentally Determined
Use Pattern o Duration o Frequency Reentry Interval	 Constant (Daily) Constant (Daily) Constant (Specific)		Distribution
Population Linked	Constant (100%)		Link to Population
Dose Factors			
Body Weight	Constant (70 kg)	➠	Distribution
Body Surface Area	Constant (21,110 cm^2)		Wt.-Dependent Variable
Respiration Rate	Constant (29 l/min)		Distribution
Dose Scaling Factor	Constant $(B. Wt)^{3/4}$		Physiologically-Based
Metabolism	No Metabolism		Pharmacokinetic Model
Risk Factors			
Benchmark Dose o RfD o ED_{10}, LED_{10}	Constant (Chemical Specific)	➠	Distribution (Chemical Specific)
Relative Potency o Q_1* o TEF's	Constant (Chemical Specific)		Distribution (Chemical Specific)
Additivity Factors			
Conditional Probability o Multi-source o Multi-chemical	1.0	➠	< 1.0
Safety Factor vs. Percentiles of Probability Distributions			
Health Standard	Use Safety Factor (e.g. 10, 100, 1000)	➠	Use Percentile of Distrib. (e.g. 95th percentile)

Exposure Factors: Diet and Water

Residue Level in Diet

Single chemical field trials conducted at maximum use rates and the shortest post-harvest interval are used in a Tier 1 analysis to establish the maximum residue of a pesticide appearing on each commodity. If no residues are found, then the tolerance is typically set at twice the limit of quantification for the analytical method. The tolerance or the anticipated residue level of the pesticide on the food commodity is used by the Dietary Residue Exposure System (DRES) program (3) to calculate the combined maximum exposure to a single chemical from all dietary sources. The chronic analysis links exposure to population, but in Tier 1 it is assumed that 100% of the entire population or subpopulation will be exposed to the pesticide at the mean value on a daily basis.

In higher tier analyses, distributional analysis of food consumption data can be obtained from the USDA food consumption data in the Continuing Survey of Intake by Individuals (CSFII) (USDA, 1989 -1992, 1994 - 1996). Physiological and demographic data such as gender, age, self-reported height and weight, ethnicity, pregnancy and lactation status, and household information permits an assessment of food consumption by specific population groups of interest. These food consumption data can be multiplied by single residue values (i.e., the mean residue value) or a distribution of measured residues or anticipated values for each food to calculate daily exposure. The residue data may come from 1) field trial studies, 2) market basket surveys conducted by the registrant, or 3) state and federal monitoring programs.

Residue Levels in Water

In the past, the EPA Office of Drinking Water allocated 20% of the reference dose of a chemical to water. The Maximum Contaminant Level (MCL) was calculated using a 100, 300 or 1000-fold safety factor depending on the quality of the hazard data and the carcinogenic classification (6). Higher tier analysis may be conducted using distributions of actual residues appearing in ground and surface water. Exposure may be linked to populations by using data collected under the Safe Drinking Water Act (10).

Market Share

Tier 1 analysis assumes that 100% of the crop is treated and residues appear at the tolerance or anticipated residue levels for all commodities having a published tolerance. In higher tier analyses, the percent of crop treated could be based on information supplied by the Office of Pesticide Programs Biological and Economic Analysis Division or by independent pesticide use surveys such as those provided by Doane Market Research, Inc. (14). Since use patterns may change from year to year, distributions of these uses can be constructed. It is expected that this rarely will be necessary.

Exposure Frequency and Duration

When chronic toxicity or oncogenicity endpoints are being considered, it is often assumed in Tier 1 risk analysis that exposure to pesticides via diet and water is not dependent upon differences between regions of the country. In higher tier analyses, however, it may be necessary to identify certain population subgroups that are not mobile and consume a regional diet or use a regional water supply throughout a large portion of their life.

In addition to chronic exposure, the Food Quality Protection Act requires that acute dietary risk analyses also be conducted with particular focus on risk to infants and children. For this purpose, exposure is expressed as a percent of the acceptable daily dose based on acute toxicity, developmental toxicity or subchronic toxicity endpoints. Tier 1 analyses may be performed for infants and children based upon tolerances. If a higher tier analysis is required, then distributional food intake and residue data may be used.

Population Linked Exposure

Tier 1 risk analysis assumes that the calculated risks are relevant to the entire U.S. population or to its subgroups, irrespective of geography. In higher tier assessments it may be necessary to determine exposure for subcomponents of the national or the regional population. Such an assessment would then be population-linked and distributions of exposure derived would be population-weighted using Monte Carlo simulation.

Exposure Factors: Non-Occupational

Turf: In Tier 1 risk analyses, exposure and internal dose are typically calculated using pesticide application rate information and default values for dislodgeable residues and dermal penetration. In higher-tier analyses, surrogate data from exposure studies conducted on other pesticides or chemical-specific information may be used to obtain average estimates of exposure and dose or distributions of these parameters. Higher tier exposure and risk analyses may also require the use of time series analysis of exposure based on a calendar-year, especially when evaluating acute or subchronic toxicity endpoints.

The Pesticide Handlers Exposure Database (PHED) has been used effectively by the EPA to characterize occupational exposure (15). Although the primary focus of the PHED database is that of the agricultural worker, the Outdoor Residential Exposure Task Force (11) is presently developing a similar, more pertinent database for outdoor residential uses. When this database is complete it will include exposure data for individuals applying pesticides to turf and subsequently re-entering the treated area, as well as for bystanders that may enter the treated area at various time intervals post-application. It is expected that the exposure information in the database will be subdivided according to application methodology/formulation type, and data quality, as has been done in the PHED database.

Indoor Residential: Exposure to pesticide formulations may be estimated either by collecting environmental samples consisting of the measurement of air and surface residues during and following an application, or the use of personal dosimetry or biological monitoring data. Various techniques for the evaluation of both primary and secondary exposures are discussed in detail in EPA's Pesticide Assessment Guideline, Subdivisions U (15) and K (16), respectively. Unfortunately, surrogate data on indoor pesticide uses have not been compiled so that conducting even a Tier 1 analysis is difficult at this time.

Higher tier exposure and risk analyses will often use time series analysis of exposure based on a calendar-year, especially when evaluating acute or subchronic toxicity endpoints. Such models provide more realistic assessment of exposures because it is unlikely that a person will be maximally exposed to more than one chemical. Monte Carlo simulation techniques are used to derive the combined exposure distributions that are weighted by the probability of exposure occurring on a given day of the year.

Penetration Factors

Tier 1 analysis typically make default assumptions about clothing barrier factors and dermal penetration. Higher tier analyses require the collection of chemical-specific data expressed either as point estimates or as distributions. In both analyses, the calculated internal dose is critically dependent on exposure duration, the degree to which the chemical penetrates protective clothing and the magnitude of dermal absorption.

Use Pattern

In Tier 1 analysis exposure is typically calculated on a daily basis and compared to an acute or subchronic toxicity endpoint. Higher tier analyses of the combined exposure to two or more chemicals requires information on market share, frequency of application, and the amount of each pesticide used on a seasonal basis, as well as details on reentry time. Generally, studies must be designed to collect residues for several days post-treatment in order to be able to estimate exposures on not only the day of application, but for subsequent days as well.

Population Linked Exposure

An evaluation of risk associated with multiple chemical use requires higher tier analyses that take into account the probability of exposure to each chemical either alone or together. In some cases it may be necessary to conduct a time series analysis of episodic events that are superimposed on a background of chronic exposure arising from agricultural uses.

Dose Factors

In Tier 1 risk analyses, the EPA generally uses a set of default values for physiological parameters that affect dose calculations (12). Sensitivity analyses can be performed to determine which, if any, of these parameters significantly impact risk calculations. When there are sufficient data available on absorption, metabolism, distribution and excretion in animals and in humans, then physiological-based pharmacokinetic models may be used to scale calculations of the internal dose in animals to man.

Risk Factors

The use of single point estimates of dose or potency in Tier 1 risk assessment have become so commonplace that these values are often treated as if they were population parameters rather than sample estimates that have distributions. In higher tier risk assessments, distributions of reference doses (RfD, ED_{10}) or cancer potency estimates may be constructed using Monte Carlo simulation techniques to calculate risk.

Additivity Factors

In Tier 1 risk analyses, it is conservatively assumed that the conditional probability of exposure to a chemical via multiple pathways is one (i.e. the consumer will be exposed daily to chemical A from all specified sources or daily to Chemical A and B in the multi-chemical scenario). Higher tier risk assessments may use market surveys to determine the conditional probability of exposure from multiple sources or to multiple chemicals.

Safety Factors vs. Percentiles of Probability Distributions

Tier 1 risk analyses typically use worst case (i.e. upper bound) estimates of exposure and worst case (i.e. upper bound) estimates of hazard or potency to arrive at the most conservative estimate of risk. For toxicity endpoints that have biological thresholds, safety factors ranging from 10 to 1000 are often utilized to take into account the quality of the scientific data as well as intraspecies and interspecies variability.

Higher tier analyses have the benefit of retaining information on the variability associated with the exposure, hazard and risk to the end of the analysis. This technique allows the risk manager to identify the probability of exposure occurring at doses less than or equal to an acceptable daily dose and the magnitude of the variability in the risk distribution for the whole population or selected subpopulations. Acceptable levels of exposure and risk therefore can be established to conform to an acceptable safety standard stated as a percentile of the risk distribution instead of using default safety factors.

AGGREGATE EXPOSURE/RISK ASSESSMENT METHODS

Criteria for Conducting Aggregate Exposure/Risk Assessments

An aggregate risk assessment for a single chemical should be conducted if:

- Exposure to the pesticide occurs via more than one source or
- Simultaneous exposure to a chemical from multiple sources is anticipated to occur in the exposure time-frame (i.e. acute, subchronic, chronic).

Criteria for Conducting Cumulative Exposure/Risk Assessments

A cumulative exposure/risk assessment should be conducted if:

- The criteria of a common mechanism of toxicity have been met and
- Simultaneous exposure to multiple chemicals is anticipated to occur in the exposure time-frame (i.e. acute, subchronic, chronic).

A common mechanism of toxicity for two or more chemicals may exist when:

- The toxic response produced by the chemicals are initiated by essentially the same sequence of major biochemical events;
- There is a common target or organ system;
- The dose response-curves based on the biologically effective dose are parallel;
- Differences in potency between chemicals are reflected as shifts in baseline of the biological effective dose-response function rather than changes in slopes; and the
- Effects of two or more chemicals are additive.

Probabilistic (Distributional) Risk Assessment

Risk assessments conducted by the EPA largely have been deterministic in nature since they have used single point estimates to characterize hazard, exposure and risk. The uncertainty introduced by using point estimates (whether an average or upper bound estimate like Q_1*) increases with the number of variables used to calculate the estimates (18). The use of distributional methods have been recommended by the National Research Council (19) for the estimation of dietary intake of pesticides and by the EPA (20) for assessments of exposures by all routes. This section illustrates how a distributional analyses can be used to address the requirements for multi-source and multi-chemical exposure and risk characterization.

Distributional analyses using Monte Carlo simulation provide a scientifically defensible methodology for combining multiple exposure sources (diet, water, non-occupational sources) arising from one or more chemicals. Briefly, the technique involves constructing probability distributions of the daily dose arising from each exposure source separately for one pesticide (Chemical A) and combining these distributions together to obtain a composite dose distribution (Figure 1). The distribution of the daily dose from exposure to Chemical A is calculated as a percentage of the RfD (Figure 1); a proportion of the RfD (i.e. Hazard Index = Exposure ÷ RfD , or 3 as a distribution of the margin of exposure where the MOE = NOEL or (ED_{10}) ÷ Exposure (Figure 2).

The distribution of this reference ratio or a percentile thereof is then judged against a safety standard in order to reach a risk management decision for Chemical A.

A similar analysis can be conducted for a second pesticide (Chemical B) and the separate probability distributions of the reference ratios for Chemicals A and B can be combined even if there are differences in relative potencies of each chemical. The composite exposure distributions for Chemicals A and B can be combined only after taking into account the probability that joint exposure to both chemicals will occur on a single day. The distribution for each chemical is then multiplied by the appropriate toxicity equivalency factor for each chemical and distributions are combined as illustrated in Figure 3.

Cumulative Risk Assessment: How to Add

Toxicity Equivalency Factors

The mathematical combination of the dose for two of more chemicals that share a common mechanism of toxicity depend on establishing a common ground for comparison. The toxicity equivalency factor adjusts the concurrent daily dose from exposure for each chemical relative to its potency according to the following equation:

$$\text{Total Dose} = (D_a \times TEF_a) + (D_b \times TEF_b) + (D_i \times TEF_i)$$

This method can be implemented by using Monte Carlo simulation to keep track of the concurrent daily dose arising from multiple pathways for two or more chemicals. These dose fractions are then multiplied by the TEF unique to each chemical to arrive at standardized doses that are then summed to give the individual's cumulative daily dose. The distribution of the individual cumulative doses can be plotted and the resulting distribution can be judged against a benchmark of acceptable risk. Information on variability is retained right up to the last step, and the risk manager can evaluate the magnitude of the overall variability and uncertainty.

Implicit in the TEF method is the assumption that the relative potency of the chemicals being compared is based on a common measure of toxicity clearly tied to the common mode of action for members of the class. If the response measures are not comparable or well understood, then the assumption of a common mechanism may be invalid. This would be true especially if it was believed necessary to apply different uncertainty factors to different chemicals in the class.

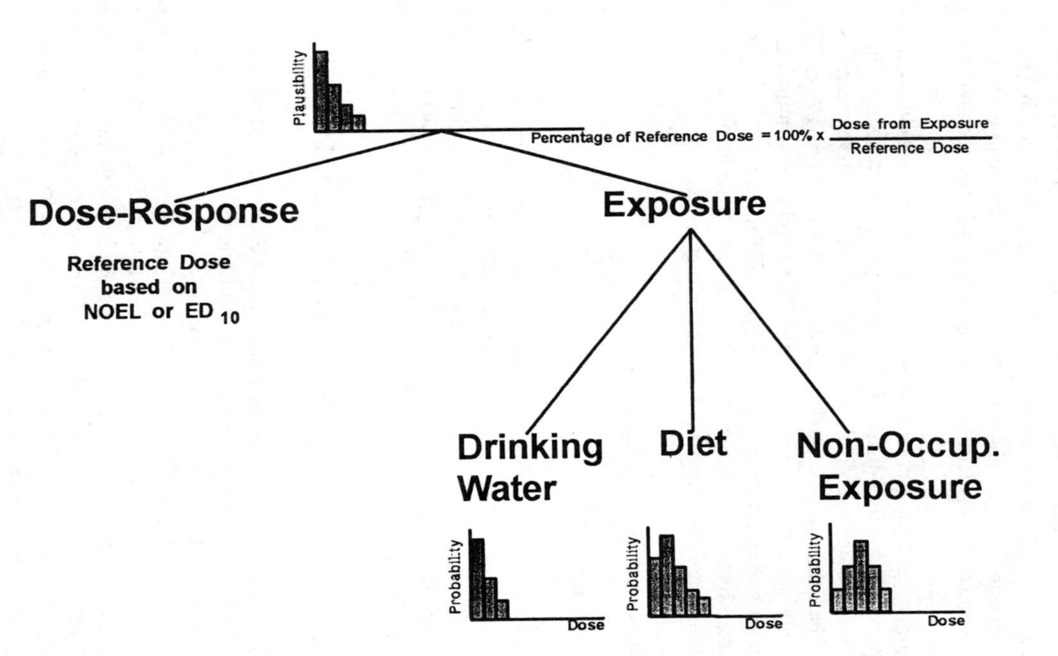

Figure 1: Chronic Toxicity Risk Characterization

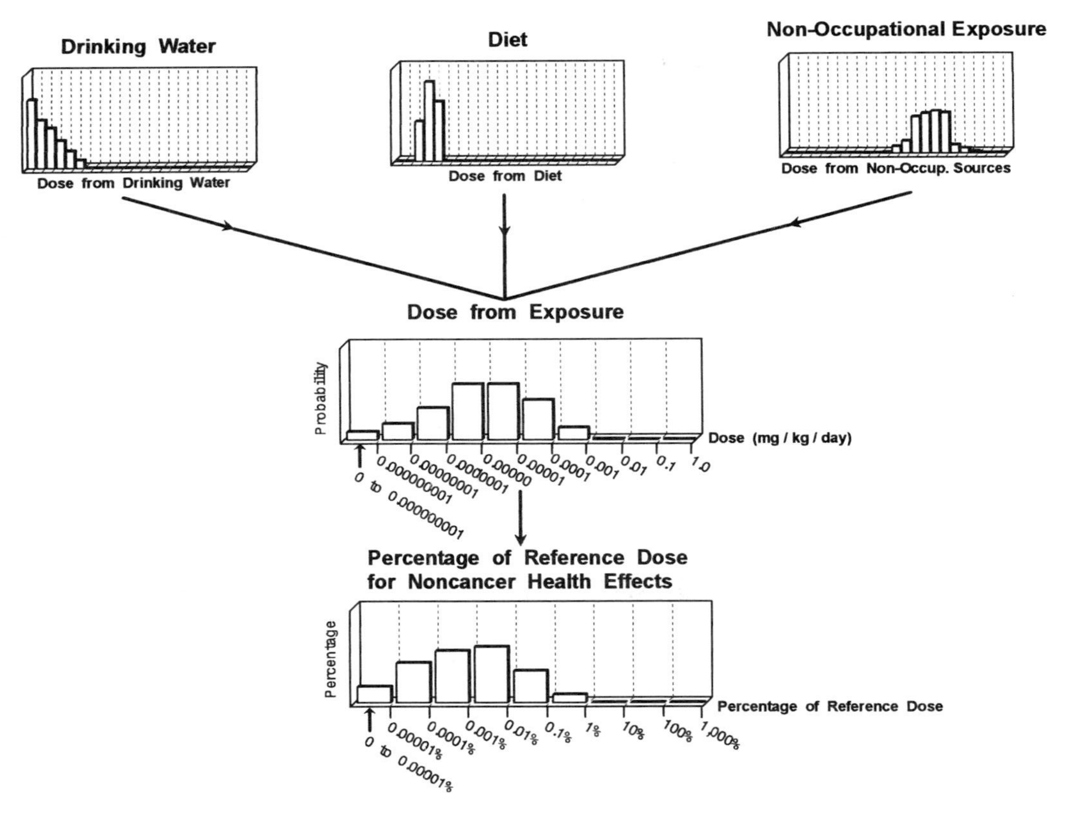

Figure 2: Aggregate Risk - Multipath

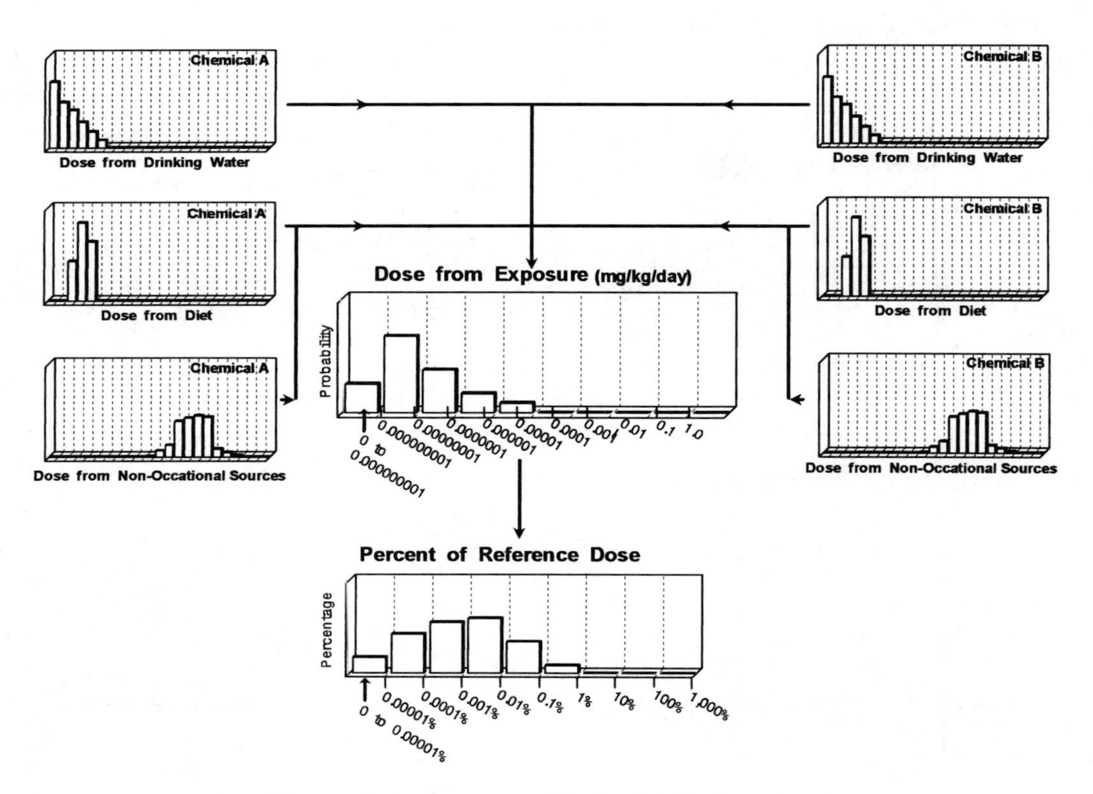

Figure 3: Aggregate Risk - Multichemical

Margin of Exposure

The margin of exposure is defined as a ratio of the NOEL or ED_{10} to the exposure. For chemicals with a common mechanism the combined MOE is defined as:

$$MOE_T = \frac{1}{(1/MOE_a) + (1/MOE_b) + (1/MOE_i)}$$

This approach has the advantage in that the MOE is defined for each chemical by its own NOEL. Ideally, because the MOE's will be added in the manner indicated above, the NOEL for compounds sharing a common mechanism will be based upon the same toxicity endpoint or biochemical surrogate, evaluated in the same species by the same route of administration and for the same duration of exposure. Experimental error in accurately defining the NOEL can be controlled by standardizing the magnitude of response across studies by using the best estimate of the ED_{10}.

Hazard Index

The method for combining Hazard Indices (HI_T) for chemical that share a common mechanism is done according to the following equation:

$$HI_T = \frac{Exposure_a}{RfD_a} + \frac{Exposure_b}{RfD_b} + \frac{Exposure_i}{RfD_i}$$

This method is acceptable as long as the RfD for each chemical is based upon similar studies and the same uncertainty factors are employed. When the studies or the uncertainty factors used in determining the RfD are different, then combining the hazard indices is not desirable because the risk manager cannot separate uncertainty from variability in the final risk distribution.

Standardization of Toxicity Endpoint Selection

The Environmental Working Group published a report (21) where they evaluated the acute dietary risk of 13 organophosphorus insecticides using references doses that were established by the EPA based on a variety of different species, and treatment durations. The data are reproduced in Table 2.

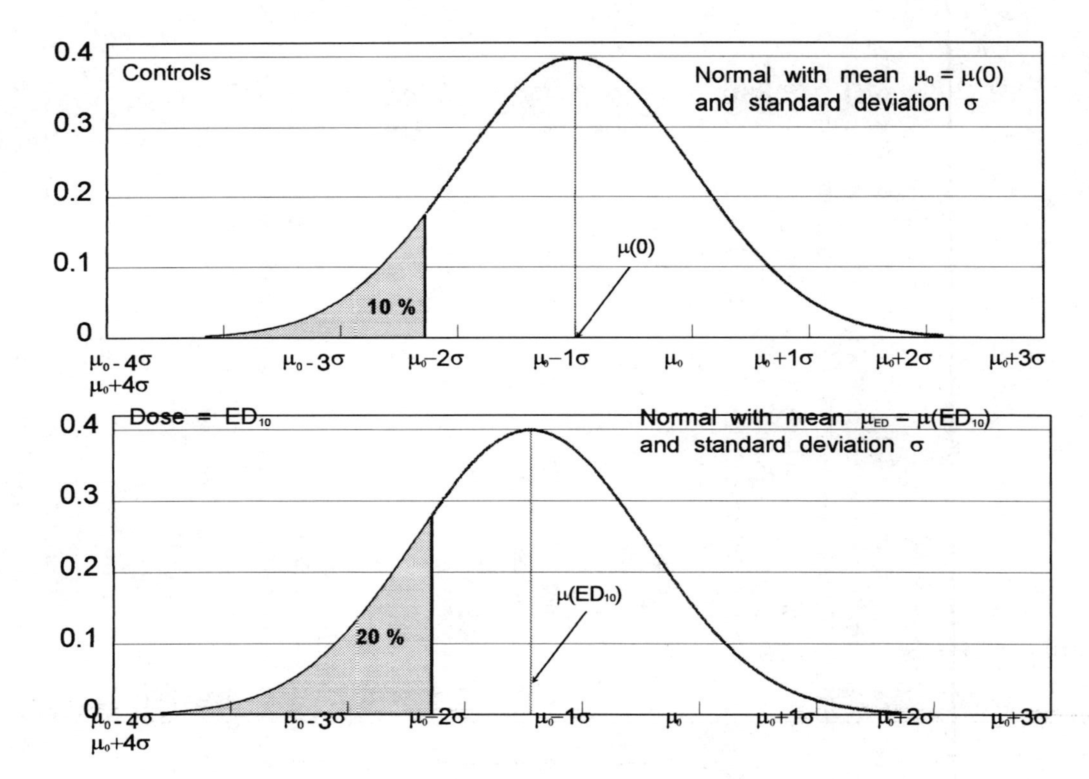

Figure 4 - ED10 ED10 defined as the dose (mg/kg) corresponding to a 10% increase in the frequency of cholinesterase levels below the 10th percentile in the distribution of cholinesterase levels in the control animals

Table 2 - Summary of Reference Doses Cited in the EWG Report (22)

Chemical	Study Type	Cholinesterase Endpoint*	Uncertainty Factor	Ref. Dose (mg/kg/day)
Acephate	90-Day Rat	P	100	0.0012
Azinphos methyl	1-Year Dog	RBC	100	0.0015
Chlorpyriphos	28-Day Human	P	100	0.0003
Diazinon	48-Day Human	P	30	0.0007
Dichlorvos	1-Year Dog	P, RBC, B	300	0.00017
Dimethoate	2-Year Rat	RBC	100	0.0005
Ethion	Human	P	100	0.0005
Malathion	2-Year Rat	P, RBC	100	0.04
Methamidophos	8-Week Rat	---	100	0.001
Methidathion	1-Year Dog	P, RBC, B	100	0.0015
Methyl Parathion	2-Year Rat	P, RBC, B	1000	0.00002
Phosmet	2-Year Rat	P, RBC, B	300	0.003
Pirimphos methyl	56-Day Human	P	3000	0.00008

* P = Plasma, RBC = Red Blood Cell, B = Brain

This approach fails to differentiate between hazard and uncertainty and inappropriately portrays the risks attributed to dietary exposure as real. As the uncertainty factors range from 30 to 3000, the reference doses are probably only weakly related to acute toxicity. In Table 3, the 13 chemicals were ranked in an increasing order according to their oral LD_{50}. Although, the correlation coefficient between the LD_{50} and the reference dose was 0.90, Figure 5 shows that there are some notable outliers where the use of a large uncertainty factor artificially skewed the chronic reference doses (e.g. primiphos methyl).

To evaluate the cumulative risk associated with organophosporus pesticide exposure, the EWG converted the reference doses for each compound to chlorpyrifos equivalents as summarized in Table 4. The use of a toxicity equivalency factor is not suitable in this case because the hazard endpoint, the study type, the species evaluated and the uncertainty factor are different for each chemical. Furthermore, because none of the studies were based on an acute toxicity measure, these benchmark doses cannot be used in an acute dietary risk assessment.

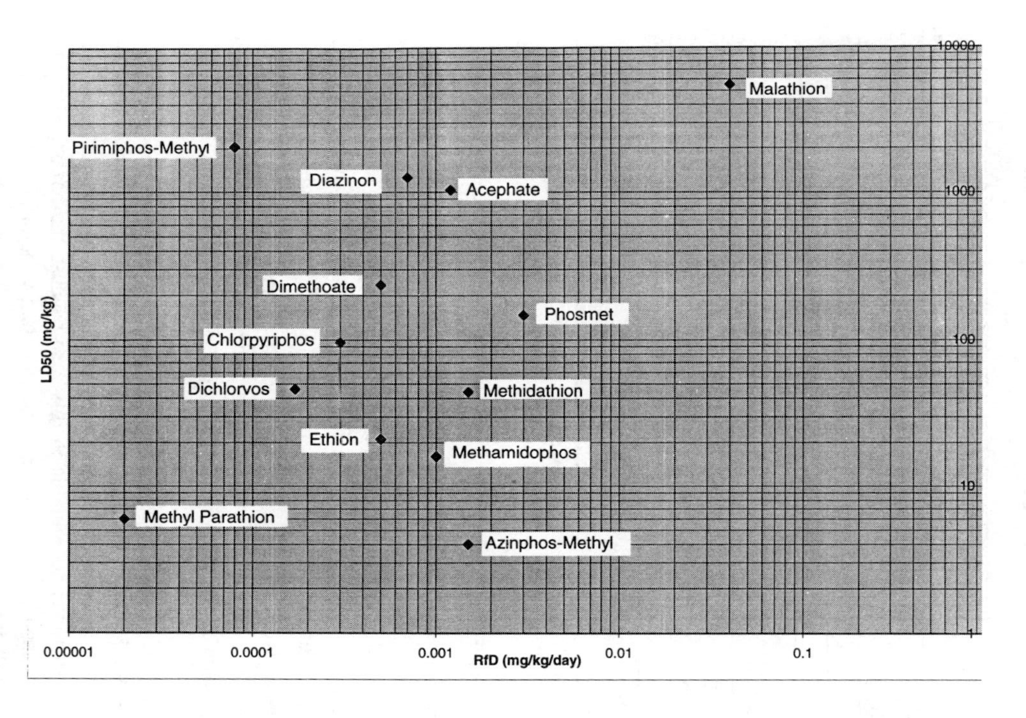

Figure 5: Scatter Plot of Oral LD50 v.s. the RfD for 13 Organophosphorus Insecticides

Table 3: Rank Order of 13 Organophosphorus Insecticides Based on the Oral LD_{50}

Chemical	Study Type	Uncertainty Factor	Ref. Dose (mg/kg/day)	LD50 (mg/kg)
Azinphos methyl	1-Year Dog	100	0.0015	4
Methyl Parathion	2-Year Rat	1000	0.00002	6
Methamidophos	8-Week Rat	100	0.001	16
Ethion	Human	100	0.0005	21
Methidathion	1-Year Dog	100	0.0015	44
Dichlorvos	1-Year Dog	300	0.00017	46
Chlorpyriphos	28-Day Human	100	0.0003	96
Phosmet	2-Year Rat	300	0.003	147
Dimethoate	2-Year Rat	100	0.0005	235
Acephate	90-Day Rat	100	0.0012	1030
Diazinon	48-Day Human	30	0.0007	1250
Pirimiphos methyl	56-Day Human	3000	0.00008	2050
Malathion	2-Year Rat	100	0.04	5500

Table 4: Chlorpyrifos Equivalent Doses[1]

Chemical	Study Type	Uncertainty Factor	Ref. Dose (mg/kg/day)	Chlorpyrifos Equivalents
Methyl Parathion	2-Year Rat	1000	0.00002	15.00
Pirimiphos methyl	56-Day Human	3000	0.00008	3.750
Dichlorvos	1-Year Dog	300	0.00017	1.765
Chlorpyriphos	28-Day Human	100	0.0003	1.000
Ethion	Human	100	0.0005	0.600
Dimethoate	2-Year Rat	100	0.0005	0.600
Diazinon	34-Day Human	30	0.0007	0.429
Methamidophos	8-Week Rat	100	0.001	0.300
Acephate	90-Day Rat	100	0.0012	0.250
Azinphos methyl	1-Year Dog	100	0.0015	0.200
Methidathion	1-Year Dog	100	0.0015	0.200
Phosmet	2-Year Rat	300	0.003	0.100
Malathion	2-Year Rat	100	0.04	0.008

In Table 5 the no observed effect levels (NOEL) derived from a repeat dose study in female Sprague-Dawley rats are compared to the NOEL obtained following a single exposure to either diazinon or methidathion. The results indicate that

(1) The NOEL based on plasma cholinesterase activity following acute exposure were 12.5 to 16.7 times greater than the NOEL derived from repeat dose studies in the same species;
(2) The NOEL based on RBC cholinesterase activity following acute exposure were 7 (methidathion) to 125 (diazinon) times greater than the corresponding NOEL from the repeat dose study;

Table 5 illustrates the importance of selecting an appropriate benchmark of toxicity when conducting a cumulative risk assessment. Table 6 shows that even when the species and sex, the response measure and the study duration are identical for two chemical, variability in the precision with which the NOEL are defined, can be problematic in a cumulative risk assessment. In Table 6, the magnitude of the response was standardized for each chemical by defining the ED_{10} as that dose that resulted in a 10% shift of the distribution of cholinesterase activity away from the mean of the control distribution (Figure 5). Using this definition, the ED_{10} fell between the statistically significant effect and the no effect level as shown in Table 7.

Table 5: Comparison of the Acute and Subchronic LD_{50} & NOEL (mg/kg) For Diazinon & Methidathion in the Female Rats

Chemical:	Diazinon		Methidathion	
Duration of Treatment:	1 Day	28 Days	1 Day	28 Days
LD_0	> 250	--	>10	
LD_{50}	1005	--	44	--
Plasma AChE	0.25	0.02	2.5	0.15
RBC AChE	2.5	0.02	1.0	0.15
Brain AChE	10	2.3	1.0	0.15
Clinical Signs/FOB	25	23	1.0	0.15

Table 6: Comparison of the Acute NOELs (mg/kg) with the ED_{10} (mg/kg) for Diazinon & Methidathion in the Female Rats

Chemical:	Diazinon		Methidathion	
Duration of Treatment:	NOEL	ED_{10}	NOEL	ED_{10}
Plasma AchE	0.25	0.6	2.5	4.1
RBC AchE	2.5	2.8	1.0	1.6
Brain AchE	10	18.7	1.0	1.0

Table 7: Comparison of the No Observed Effect Level to the ED_{10} for Diazinon

Male Rats				Female Rats			
Dose (mg/kg)	Mean Cholinesterase Activity (Mmol/ml)			Dose (mg/kg)	Mean Cholinesterase Activity (Mmol/ml)		
	Plasma	RBC	Brain		Plasma	RBC	Brain
0	249	944	899	0	991	1224	749
0.05	253	967	938	0.05	738	1014	699
0.5	232	956	738	0.12	1070	1084	832
1.0	255	968	744	0.25	855	1131	694
10	$ED_{10} = 4.7$ 139*	$ED_{10} = 7.7$ 792*	755	2.5	$ED_{10} = 0.6$ 390*	1072	766
100	74*	478*	910	25	146*	$ED_{10} = 2.8$ 793*	$ED_{10} = 18.7$ 477*
500	26*	606*	$ED_{10} = 319$ 278*	250	22*	676*	228*

* Statistically significantly different from the control group ($p \leq 0.05$)

AN AGGREGATE EXPOSURE ASSESSMENT EXAMPLE

EPA has proposed using environmental fate models as a preliminary screen to predict the maximum amounts of a pesticide that might reach ground (7) or surface water (8, 22) following agricultural use. A higher tier model has been developed to predict the concentration of pesticide residues that might appear in surface water residues based upon monitoring data (9). A higher tier distributional analysis of residues of the herbicide atrazine in ground and surface water has been conducted (10) using drinking water monitoring data collected under the Safe Drinking Water Act (23). For the purposes of illustration, the results from multiple tiers are compared in Table 8 and the Tier 3 distributional analyses are presented graphically in Figures 6 & 7. The results indicate that Tier 1 environmental fate models are extremely conservative. Exposure assessment based upon monitoring data for a specific chemical are more realistic and have been validated in Tier 3 analyses based on annual monitoring data collected under the Safe Drinking Water Act.

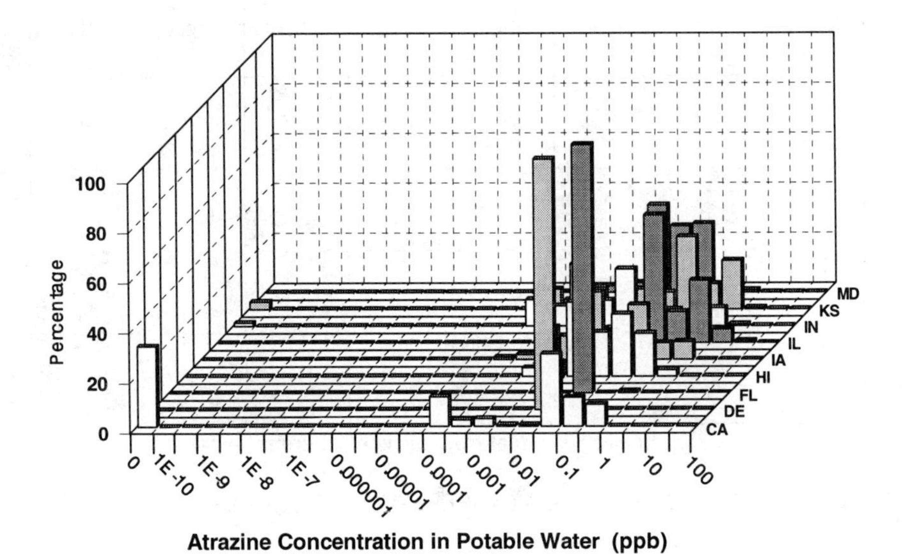

Figure 6: Distributional Characterization of Atrazine Concentration (ppb) in Drinking Water of 9 of the 18 Major Use States

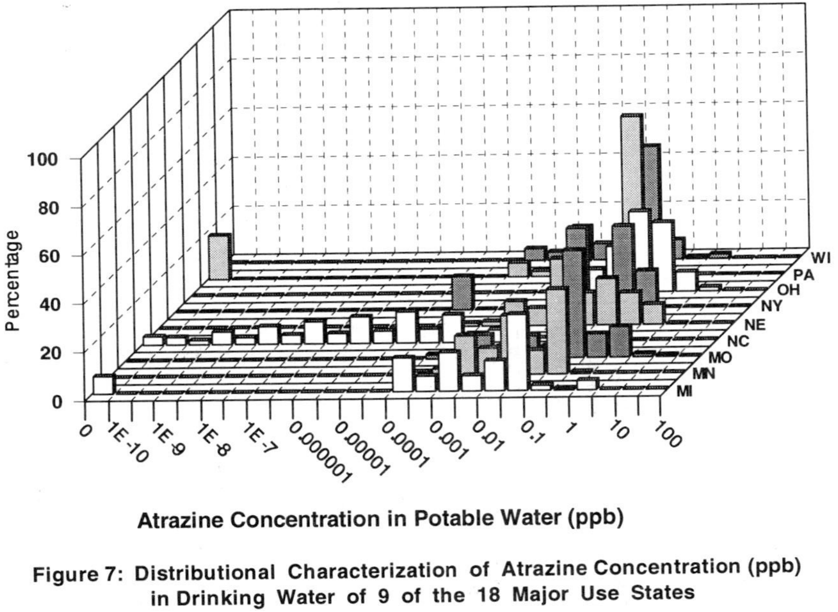

Figure 7: Distributional Characterization of Atrazine Concentration (ppb) in Drinking Water of 9 of the 18 Major Use States

Table 8: Tier 1, Tier 2 and Tier 3 Estimates of Atrazine Concentration (ppb) in Ground and Surface Water** following Use on Corn

Tier Assessment	Surface Water	Ground Water
1 Screening Models	GENEEC* Peak Concentration = 108 ppb Day 56 Conc. = 97.7 ppb	SCI-GROW 7.9 ppb
2 Model based on Monitoring Data	SWMI 95th Percentile = 10.9 ppb	----
3 Population-Linked Monitoring Data for 21 States	% Non-Detects = 69%*** No. Assessed = 75.6 MM 95th Percentile = 2.1 ppb	% Non-Detects = 97%*** No. Assessed = 49.2 MM 95th Percentile = 0.5 ppb

*Assumes 2.1 pounds ai applied to corn either pre-emergence or as a split application.
**Surface water includes blended water or water from unknown sources.
***Non-detects were substituted at one-half the limit of detection.

Figure 6 and 7 present distributional data for atrazine residues in drinking water sources for 21 major use states (24). These data illustrate the considerable variability in concentrations not apparent when point estimates or 95th percentiles of distributions are used. This approach has the additional advantage of being able to weight the results by population density and to identify subpopulations where it may be appropriate to implement mitigation measures.

Discussion and Conclusions

The Food Quality Protection Act has focused attention on methodology for assessing the aggregate risk associated with exposure to a single chemical from multiple sources (i.e. diet, water and non-occupational sources) and to multiple chemicals that have a common mechanism of toxicity. This document outlines a risk assessment strategy that conserves resources and expedites regulatory decision making for simple cases, but is scientifically rigorous enough to deal with complex cases.

This paper identifies significant sources of exposure to pesticides and provides a decision logic for determining when additional exposure data is needed. A three-tier assessment approach is proposed; the lowest tier uses default assumptions and single point (deterministic) estimates of exposure, hazard and risk are employed. Tier 2 assessments utilize a combination of deterministic and probabilistic (distributional) data while Tier 3 assessments rely predominantly on distributional data.

The use of probability distributions to characterize exposure and risk make it possible to more realistically combine:

- multiple years;
- multiple subpopulations;
- multiple exposure sources; and
- multiple chemicals

Worst case, relatively improbable values do not have to be taken as representing reality for each component. All of the information on each component of the hazard and exposure assessment are carried through to the end instead of requiring interim single number characterizations at different stages of assessment.

Probability distributions can and do reflect:

- the continuum of data from the smallest to the largest value of a component;
- the relative likelihood of each of the values in that range;
- the uncertainty associated with the measurement of each component; and
- the variability from one individual to another in the population.

By using distributional data for decision making, the risk manager can

- identify the probability of exposure occurring at doses less than or equal to an acceptable daily dose;
- define the magnitude of the variability in the risk distribution for the whole population or selected subpopulations; and
- ensure that the proposed levels of exposure and risk conform to an acceptable safety standard stated as a percentile of the risk distribution.

Additional work is still needed on the following tasks:

- Develop guidelines for the use of default values in risk assessments;
- Develop guidelines for selecting surrogate data when information is missing;
- Evaluate alternate methods for conducting risk assessment, including methods to estimate the probability of co-exposure to pesticides from multiple sources;
- Enhance or develop reliable exposure data needed for quantifying exposure to pesticides in the diet or in drinking water and from non-occupational sources;
- Develop methods that more accurately link exposure to specific sub-populations; and
- Facilitate the training of scientists and risk managers in the conduct and the evaluation of probabilistic risk assessments.

References

1. Bliley, R. 1996. *Food Quality Protection Act of 1996.* 104 Congress, 2nd Session. Report 104-669, part 2, pp. 1-89. Washington; D. C: Government Printing Office.
2. U.S. Environmental Protection Agency (EPA). 1992. Guidelines for Exposure Assessment. Final guidelines for exposure assessment. May 29,1992. *Fed. Reg.* 57(104): 22888-22937.
3. Petersen, B., and Chaisson, C. F. 1988. Pesticides and Residues in Food. *Food Technology* 42(7): 59-64.
4. U.S. Environmental Protection Agency (EPA). 1996, Final Office Policy for Performing Acute Dietary Exposure Assessment. Office of Pesticide Programs.
5. United States Department Of Agriculture (USDA), 1998. Continuing Survey of Food Intake by Individuals, 1994-96. Riverdale, Maryland.
6. U.S. Environmental Protection Agency (EPA). 1991. National Primary Drinking Water Regulations: Final Rule. Part II. January 20, 1991. *Fed. Reg.* pp.3525 - 3757.
7. Barrett, M., Initial Tier Screening of Pesticides for Ground Water Concentrations Using the SCI-GROW Model. United States Environmental Protection Agency, December 3, 1997.
8. Parker, R. D., Nelson, H. P., Jones, R. D., & Mostaghimi, S. Development for Screening Level Estimation of Pesticide Exposure in the Aquatic Environment. United States Environmental Protection Agency. Washington, DC.
9. Chen, W., Hertl, P., & Tierney, D. A Simple Regression Model for Predicting Surface Water Concentrations Resulting from Agricultural Field Runoff and Erosion., Novartis Crop Protection, Inc., 1998.
10. Sielken R.L., Bretzlaff, R.S. and Valdez-Flores. 1998. Probabilistic Risk Assessment Using Atrazine and Simazine as a Model. Triazine Herbicides: Risk Assessment. American Chemical Society Symposium Series # 683, Washington, DC, 1998.
11. Selman, F. Outdoor Residential Exposure Task Force (ORETF). 1998. Personal Communication.
12. U.S. Environmental Protection Agency (EPA).1997. *Exposure Factors Handbook. General Factors.* U.S. Environmental Protection Agency, National Center For Environmental Assessment, Washington D.C. EPA/600/P-95/002Fa.
13. American Industrial Health Council, Exposure Factors Sourcebook, Washington, DC, 1994.
14. Doane Marketing Research Inc. St. Louis, MO 63146
15. Pesticide Handlers Exposure Data Base (PHED). 1992. Notice of Availability of the Pesticide Handlers Exposure Data Base through Versar, Inc. 3 June 1992. *Fed. Reg.* 57(107): 23403.

16. U.S. Environmental Protection Agency (EPA).1987. Pesticide assessment guidelines for applicator exposure monitoring - Subdivision U. Office of Pesticide Programs. Office of Pesticides Programs. Office of Pesticide and Toxic Substances. Washington, D.C. EPA-540/9-87/127.
17. U.S. Environmental Protection Agency (EPA). 1987. Pesticide assessment guidelines -Subdivision K. Office of Pesticide Programs. Office of Pesticides Programs. Office of Pesticide and Toxic Substances. Washington, D.C.
18. Finley, B. L. and Paustenbach, D.J. 1994. The Benefits of Probabilistic Exposure Assessment: Three Case Studies Involving Contaminated Air, Water, and Soil. *Risk Anal.* 14: 53-60
19. National Research Council (NRC). 1993. *Pesticides in the diets of Infants and Children.* National Academy Press, Washington, D.C.
20. Guidance for Submission of Probabilistic Exposure Assessments to the Office of Pesticides Programs, Health Effects Division. United States Environmental Protection Agency,1998.
21. Environmental Working Group. 1998. *Overexposed: Organophosphate Insecticides in Children's Food.* FIFRA Scientific Advisory Panel (SAP) Meeting.
22. Hetrick, J. 1998. *Proposed Methods for Basin-scale Estimation of Pesticide Concentrations in Flowing Water and Reservoirs for Tolerance Reassessment.* FIFRA Scientific Advisory Panel (SAP) Meeting. July 29-30, 1998
23. Safe Drinking Water Act (SDWA) Federal Register, 40 CFR Parts 22 & 142 January, 1991. 3752; Amended August 6th, 1996.
24. Sielken R.L., Bretzlaff, R.S. and Valdez-Flores. 1998.Risk Characterization for Atrazine and Simazine. EPA MRID No. 43934415.

Chapter 6

New Science, New Processes, and New Problems: The Food Quality Protection Act from a State Perspective

Jean-Mari Peltier

Chief Deputy Director, California Department of Pesticide Regulation, 830 K Street, Room 307, Sacramento, CA 95814

When Congress passed the Food Quality Protection Act (FQPA) in August 1996, it dramatically changed how pesticides are regulated in the United States. This overview describes FQPA's impact on how we assess risk from dietary exposure to pesticides and, by extension, how we manage that risk. These responsibilities underscore the need for new data and new analytical tools that can work effectively in the post-FQPA environment.

At the outset, it should be emphasized that no criticisms are aimed at the goals of FQPA, or at the staffers of the U.S. Environmental Protection Agency (U.S. EPA), who are struggling mightily to keep FQPA from crushing the federal pesticide regulatory machinery. However, there is a need for critical examination of policy decisions being made without regard to the flexibility Congress intended for FQPA's critical start-up years.

BACKGROUND -- California Agriculture and the California Department of Pesticide Regulation

California has had a pesticide regulatory program for nearly one hundred years. Our citizens--through their Legislature--have established a comprehensive body of law to control every aspect of pesticide sales and use and to assure that we also have the tools to assess the impacts of that use. The first pesticide-related law was passed in this state just after the turn of the century, and since the 1960s, a whole body of modern, increasingly science-based pesticide law and regulation has emerged.

But California's regulatory program is more than a matter of law. It is a social imperative. California agriculture is a critical part of the state's economy. To address the needs of a large and diverse agriculture while continuing to protect public health and the environment requires a program designed and maintained for California's unique conditions.

To provide some perspective: California farmers outproduce every state in the nation. They grow more than 250 different crops and livestock commodities, yet no one crop dominates the state's agricultural economy. California leads the U.S. in production of more than 75 crops, and is the exclusive U.S. commercial producer of almonds, artichokes, dates, figs, kiwifruit, olives, persimmons, pistachios, prunes, raisins, and walnuts.

In 1996, California produced nearly 14 million tons of fruits and nuts and 20 million tons of vegetables--accounting for more than half of U.S. production. California farmers and ranchers recorded cash farm receipts of almost $24.5 billion in 1996. This on-farm income generated more than $70 billion in total economic activity throughout the state--about 9.5% of the state's total income. Nearly one-third of California's 100 million acres are devoted to agricultural production.

With this volume and variety of agriculture--much of it focused on high-value, high-labor fruit, nut, and vegetable crops with their attendant pesticide use--it is not surprising that we believe in a strong pesticide regulatory system. A strong program protects the public and the food industry by providing accountability and credibility. The program, which evolved under the Department of Food and Agriculture, received departmental status in 1991 when Governor Wilson created the California Environmental Protection Agency.

The Department of Pesticide Regulation (DPR) holds primary responsibility for regulating all aspects of pesticide sales and use to protect the public health and environment. Our mission is to evaluate and mitigate impacts of pesticide use, maintain the safety of the pesticide workplace, ensure product effectiveness, and encourage the development and use of reduced-risk pest control practices while recognizing the need for pest management in a healthy economy.

DPR's strict oversight begins with product evaluation and registration, and continues through statewide licensing of commercial applicators, dealers and consultants, local permitting and use enforcement, environmental monitoring, and residue testing of fresh produce. The Department has an annual budget of $45 million and a staff of about 400 persons--about a quarter of them scientists-- including more than 30 toxicologists and more than 50 environmental scientists. Their work is augmented by approximately 325 biologists working on local pesticide enforcement for agricultural commissioners in the state's 58 counties.

We have always been regulatory pioneers:

- California's first law regarding pesticides was passed in 1901, nine years before the passage of the first federal legislation over pesticides. The law, limited to one pesticide known as "Paris Green," an arsenical compound, dealt only with product quality and consumer fraud.

- In 1910, Congress passed the Federal Insecticide Act. The California Legislature followed in 1911 with passage of a similar bill. Both laws primarily addressed mislabeling and adulteration.

- In 1921, California took the lead with legislation that required registration of pesticides before sale in California. This law gave the state authority to cancel or deny registration of products found ineffective, or harmful to health or the environment. (It would be another quarter-century before FIFRA gave federal officials the right of premarket clearance of pesticide products.)

- We began analyzing produce for pesticide residues in 1926. This monitoring program was designed not only to safeguard the consumer against harmful residue levels, but also to ensure that no shipments of California fruit were confiscated because of excess residues.

- In the 1950's, California began asking some users of agricultural pesticides--primarily those in the business of applying pesticides--to report to the state the amount of pesticides they used, and on what crop. In 1990, California became the first state to require full reporting of all agricultural pesticide use, a program made possible by California's unique system of county agricultural commissioners.

- In the 1980's, increasing concerns about possible adverse effects of pesticides led to the passage of legislation in California that required that chronic health effects data on all pesticides be brought up to current standards. The state's pesticide regulators were charged with analyzing the data and canceling any pesticide with adverse effects that could not be mitigated. This led to the creation of a separate Medical Toxicology Branch to evaluate toxicological data and conduct risk characterizations.

DPR's scientific and technical expertise has won a reputation for excellence, and it's a standard we strive to maintain. California is the only state with a regulatory program that evaluates toxicology and other data required for pesticide registration. Our program also conducts comprehensive risk assessments, including assessment of dietary risk. We believe that our decisions must be based on the best science available. Our scientists work with U.S. EPA in many areas to develop that science.

DPR AND THE FOOD QUALITY PROTECTION ACT

DPR and Harmonization with U.S. EPA

In 1994, the Department began a program to harmonize its pesticide registration more closely with U.S. EPA. In March 1995, the two agencies signed a memorandum of understanding to more closely coordinate the federal and California registration programs. Harmonization goals include reducing needless duplication, getting safer products to market faster, and more quickly removing products that pose unacceptable hazards. Resources saved can be spent on accelerating the registration of lower-risk products. The long-term objective is to resolve differences and increase uniformity.

Harmonization brought the two agencies closer together, sharing data reviews and building solid working relationships between scientific staffs. However, FQPA diverted U.S. EPA's attention from harmonization; at the same time, FQPA made harmonizing with U.S. EPA much more vital from DPR's view. The FQPA workload at U.S. EPA, combined with a rolling reorganization at the Agency, slowed progress on harmonization. At the same time, DPR staff joined several FQPA working committees and informally participated in NAFTA harmonization talks. DPR's post-FQPA goals are to resolve problems before they require harmonization; generate a genuine work-sharing relationship, and create a harmonious process where one agency may freely use the work products of another.

OVERVIEW OF FQPA

FQPA substantially changed how U.S. EPA regulates pesticides. Among its major provisions are:

- **A new safety standard for all pesticide residues in food**
  - Requires "Reasonable Certainty of No Harm" from exposure to residues;
  - Requires consideration of aggregate assessment of all non-occupational sources of exposure, including drinking water, residential, and dietary exposure;
  - Requires assessment of cumulative exposure to a pesticide and other substances with common mechanisms of toxicity.

- **Special protections for infants and children**
  - Consideration of children's special sensitivity and exposure to pesticides;
  - Use of an extra safety factor of up to 10-fold, in addition to the traditional 100-fold safety factor;
  - Explicit determination that a tolerance (legal residue limit) is safe for children.

- **Tolerance assessment and reassessment**
  - Application of new safety standard to all tolerances issued after August 3, 1996;
  - Reassessment, within 10 years, of all tolerances issued before enactment of FQPA to ensure they meet the new safety standard;
  - Establishment of tolerances for emergency exemptions issued under Section 18 of the Federal Insecticide, Fungicide, and Rodenticide Act (FIFRA).

FQPA: MAJOR IMPACTS IN PROCESS AND SCIENCE

FQPA implementation has resulted in significant impacts on both the regulatory and scientific decision-making processes.

Regulatory impacts: FQPA changed the process for evaluating Section 18 applications. (Section 18 of FIFRA allows U.S. EPA to exempt a state from pesticide use regulations if the Agency finds that emergency pest conditions exist. The exemption allows use of a pesticide that has not been registered under FIFRA. Section 18s are particularly important in states such as California, where fruit, nut, and vegetables are considered "minor" markets compared to "major" nationwide acreage in wheat and corn). Under FQPA, the agency must establish maximum pesticide residue limits, or tolerances, for uses authorized under Section 18. U.S. EPA's interpretation of the new requirement has been to trigger a full-scale reassessment for all tolerances--not just the Section 18 crop--even though the Section 18 affects only one commodity for a limited time.

This interpretation causes significant delay and sometimes prevents issuance of a Section 18, even when the incremental risk is negligible or non-existent. This has forced DPR to issue more crisis exemptions, which are Section 18s for which no tolerance is set when a pesticide is applied, EPA has allowed states to issue more crisis exemptions (provided they believe a tolerance can be established by harvest) because they have been unable to complete their reviews in a timely manner. From 1985 to 1995, California averaged fewer than four crisis exemptions per year. In 1996, three of the six crisis exemptions issued by DPR resulted from FQPA. In 1997, FQPA crisis exemptions jumped to 20 (out of 23 total crisis exemptions) in California.

In an attempt to speed the Section 18 process in California, DPR has diverted staff to conduct Section 18 risk assessments. These include reviews of a pesticide's toxicology; potential for worker, dietary, and other exposures; and other data. DPR's risk assessment expertise makes our program uniquely qualified to assist U.S. EPA, and we have been working closely with the Agency to establish "assumptions" on which to base an assessment. While DPR wants to expedite whatever process U.S. EPA dictates, it would be preferable if U.S. EPA established a simpler procedure for emergency tolerances under FQPA. Indeed, the statute offers U.S. EPA the flexibility to adopt a different approach to Section 18 tolerances, as Congress intended.

One alternative put forth by an advisory committee to U.S. EPA on which DPR participated would recognize the temporary nature of Section 18 uses. This alternative process would focus on the incremental risk associated with the Section 18 use itself, as opposed to the total risks associated with all uses of an active ingredient. If the incremental risk calculated for the Section 18 use is insignificant--defined as less than 1 percent of the acceptable risk associated with the appropriate toxicological endpoints--then the use could be approved without a complete risk assessment. However, if the risk is deemed greater than the 1 percent cutoff, then an aggregate risk assessment would be required. This alternative approach would be fully protective of public health, while expediting Section 18s to the benefit of U.S. EPA, registrants, and growers. U.S. EPA's response has not been encouraging, but we continue to press our case.

Science impacts: We seek to continue harmonization with the process still in flux, coming to grips with U.S. EPA's evolving notion of "reasonable certainty of no harm" and new mandates to consider aggregate and cumulative effects.

Aggregate exposure assessment: California has been considering multiple routes of exposure in its assessments for several years. In some cases, DPR generated its own studies for assessing workplace and indoor exposure. In most of these cases, the route of exposure driving the risk assessment has been non-dietary; dietary exposure was considered only in the context of adding risk. For example, in assessing the risk to farm workers from occupational exposure, we added potential dietary contributions, but our main focus was workplace exposure. Under FQPA, workplace exposure is not included in aggregate exposure; U.S. EPA must evaluate occupational exposure during reregistration of the pesticide. Whether exposure is dietary or non-dietary, we should concentrate on obtaining realistic data.

Cumulative exposure to chemicals with a common mechanism of toxicity:
The need to consider this issue was highlighted in the National Academy of Sciences 1993 evaluation of the methods the federal government uses to estimate the health risks to infants and children from dietary exposure to pesticide residues. Early last year, U.S. EPA proposed an approach that assumes a common mechanism of toxicity where pesticides show a common toxicological endpoint and structural similarity.

The International Life Sciences Institute (ILSI) is working with U.S. EPA to better define what constitutes a common mechanism of toxicity and how to conduct cumulative risk assessments. ILSI has completed an initial study to define common mechanism of toxicity, using organophosphate pesticides as a case study. They have concluded that this group of chemicals share a common mechanism of inhibiting acetylcholinesterase. They reached this default conclusion because there was insufficient data to distinguish subgroups of organophosphates, which produce a variety of clinical signs not identical to all compounds within this group. U.S. EPA has used this conclusion to develop different scenarios for regulatory action. Proposing to revoke all authorized uses of these chemicals - sooner rather than later - was one scenario floated by U.S. EPA (although the agency later denied it was considering such broad-brush actions).

The science to determine common mechanism of toxicity is still being defined. Our concern is that decisions will be made to default to larger, less meaningful groupings, rather than wait for data to devise realistic subgroupings.

"The Risk Cup": Facing a lack of data for certain dietary and non-dietary exposures, U.S. EPA created the concept of the "risk cup" as an interim strategy for fulfilling the mandates of the FQPA. (In January 1997, U.S. EPA published an interim decision logic to address FQPA risk assessment issues. As defined by U.S. EPA, "the risk cup" logic is based on the concept that the total level of acceptable risk from a pesticide is represented by the pesticide's Reference Dose [RfD]. This is the level of exposure to a specific pesticide that a person could receive every day for seventy years without significant risk of a long-term or chronic non-cancer health effect. The analogy of a "risk cup" is being used to describe aggregate exposure estimates. The full cup represents the total RfD and each use of the pesticide contributes a specific amount of exposure that adds a finite amount of risk to the cup. As long as the cup remains unfilled, meaning that the combined total of all estimated sources of exposure to the pesticide has not reached 100 percent of the RfD, U.S. EPA can consider registering additional uses and setting new tolerances. If it is shown that the risk cup is full, U.S. EPA has taken the position that no new uses could be approved until the risk level is lowered. This can be done by the registrant providing new data which more accurately represent the risk or by implementing risk mitigation measures. While this explanation is focused on chronic non-cancer risk, the agency will use a similar logic to assess acute risk and cancer risk.)

U.S. EPA has proposed filling the risk cup with very conservative assumptions for drinking water and non-dietary indoor and lawn and garden exposures to pesticides. There is little data on these drinking water and non-dietary exposures, and in the absence of data, U.S. EPA assumes that 10 percent of the risk cup is given over to drinking water and 10 percent to home and garden uses. This has a great impact on the many uses of a pesticide--or in the case of organophosphates, a whole group of pesticides--on hundreds of fruits and vegetables.

However, a recent experience at DPR indicated that even this conservative interim decision logic can be preferable to decisions based on inadequate data and inappropriate use of modeling. Earlier this year, U.S. EPA was poised to deny a Section 18 application from California to use the fungicide maneb on walnuts. U.S. EPA said its calculations showed that the maneb risk cup was full, based on assumptions about exposure to dietary water residues of ETU, a breakdown product of maneb and other chemically related fungicides.

In this instance, U.S. EPA did not use the 10 percent default assumption--it used actual data and then made some wildly inappropriate calculations based on that data. Nationwide, more than 1,500 wells were sampled for ETU. One valid quantifiable detection was found in Illinois. The concentration was high--16 ppb. Yet ETU was not found in other wells monitored in the same county. DPR's experience with pesticide ground water monitoring suggests that a single detection at a relatively high level indicates a point source contaminant, as opposed to normal agricultural use. Point source contamination could be caused by proximity of a mixer/loader site. In addition, ETU is widely used as an accelerant in neoprene and other rubber production.

DPR has never had a verified detection of ETU in thousands of well samples taken in California. In Florida, where this class of chemicals is widely used, no confirmed ETU residues have been found. Nonetheless, U.S. EPA used this single, questionable detection--from among 1,500 wells--to characterize the nation's entire drinking water supply. Moreover, U.S. EPA used a new empirical screening model never intended for use as a decision endpoint. The good news is that U.S. EPA ultimately allowed a Section 18 for maneb, using the previous year's unexpired tolerance.

CRITICAL SCIENTIFIC ISSUES

Since FQPA took effect, U.S. EPA has presented several papers on a number of scientific issues at numerous meetings of its Scientific Advisory Panel (SAP). However, U.S. EPA's changing use of different default assumptions and different sources of data has been a source of confusion and frustration. From DPR's perspective, it appears that U.S. EPA staff sometimes have reached for very conservative default assumptions or single "worst-case" data points in aggregate exposure analysis. DPR cannot discern why such values were selected.

Distinction between screening and refined risk assessment: We believe estimating risk based on many interim default assumptions in the absence of data should be viewed as a screening assessment, rather than a realistic and conclusive risk assessment. A margin of exposure (MOE) derived from a screening assessment should not be used to draw a bright line on the "risk cup." Since U.S. EPA is collecting all existing data for addressing the FQPA issues, we hope that the final "interim" policies on risk assessment will not call for injudiciously stacking many high-end defaults to address aggregate and cumulative exposures. Risk managers must be able to consider the uncertainty associated with risk estimates when numerous default assumptions are used in risk assessments.

Data development: The default assumption problem underscores the need for better exposure data, especially for complex situations involving multiple chemicals or multiple routes of exposure. There is also a critical need for geographically-specific data. California registrants have developed exposure data at DPR's direction. In some cases, DPR developed its own data. However, this is an area where U.S. EPA leadership would be welcome. For example, user groups are concerned that if many active ingredient uses are dropped after U.S. EPA's interim assessment, there will be no incentive for registrants to develop actual data. U.S. EPA needs data from studies that are statistically designed and provide toxicologically pertinent detection levels. The Agency also should take into account all media which are major contributors to an exposure (e.g., foods consumed frequently and/or high in residues; data for indoor use). Data from food should be as close to the point of consumption as possible. Exposure assessments based on such data will provide a more realistic perspective for U.S. EPA.

When utilizing data that is not generated by registrants, U.S. EPA's Office of Pollution Prevention (OPP) needs to verify that its information is accurate and up to date. This suggests a role for both pesticide users and government agencies. DPR has required full reporting of all agricultural pesticide use since 1990. Recently, DPR has utilized this wealth of information to develop a database of pest management strategies that may be affected by upcoming tolerance reassessment.

Such analyses have allowed DPR to initiate proactive programs. For example, we established Pest Management Alliance grants to create partnerships targeted at reducing pesticide risks to workers, consumers, and the environment. Alliance proposals are evaluated in light of the most critical pest management needs with widespread implications. Grant recipients also must agree to match DPR funds of up to $100,000, with an initial $5,000 award to gather pest management information. The projects themselves may employ applied research, demonstration projects, or a combination of the two.

The benefits are immediate and tangible. First, DPR wants to focus the user community on research needed to move beyond FQPA. Second, we want to provide information to U.S. EPA on how materials are used, both on their own merits and as substitutes for other, threatened materials. Some of these alternatives may present non-dietary problems that are not the focus of FQPA (e.g., worker exposure). Other alternatives might be potential environmental contaminants. Whatever the case, it is critical that U.S. EPA fully understand the impact of its decisions.

To obtain the best data, U.S. EPA also needs to expand its perspective beyond the registrant and user communities. The Agency might begin by working with its pesticide and water programs to obtain data from the state agencies that monitor water. As part of that process, U.S. EPA could take into account the success of state programs in mitigating water problems. The Agency needs to quickly proceed with gathering this information

Without such efforts, FQPA will require the use of extremely conservative assumptions: for example, a commodity would presumably contain residues for all chemicals registered for use on that commodity. Such an assumption is obviously unrealistic. The complexity of these issues would escalate when the cumulative exposures also included aggregate exposures from each pesticide. Absent careful considerations, the stacking of such conservative assumptions grossly distorts the risk assessment.

To move away from such assumptions, U.S. EPA needs data on the profiles of pesticide use, coupled with data on the coexistence of pesticides in foods. U.S. EPA should make every attempt to utilize all available data. For example, we understand that U.S. EPA may not use actual (marketbasket) residue data for acute dietary assessment if the data is from co-mingled samples. However, in DPR's experience, there are very few commodity-chemical combinations where co-mingling could significantly alter the result of the assessment. Good minds at OPP should explore this issue; otherwise, a significant data base will be ignored. This is another example where U.S. EPA could benefit from expertise at the state level.

Data analysis tools needed: To properly implement FQPA--and to fulfill the goals of the National Academy of Sciences report that led to passage of the law--will require new data analysis tools. We will need valid methods of distributional (probablistic) analysis to avoid unrealistic use of multiple high-end exposures. The Monte Carlo method is the most commonly used of these.

In the conventional point estimate approach, the exposure is expressed as a fixed value, usually at the high-end of an exposure spectrum. The aggregate and cumulative exposures are then calculated by adding the exposures from all routes and all chemicals, leading to unrealistic exposure estimates. It is extremely unlikely that any individual would be exposed at the high end of all routes of exposure to all chemicals.

Alternatively, the distributional approach takes into account the entire distribution of each exposure component (e.g., consumption and residues in a dietary exposure assessment, or dietary and residential in an aggregate exposure assessment) and arrives at a final distribution of exposures that would most likely demonstrate that the probability of having all routes of exposures all at the high-end in a highly improbable range that is statistically unreliable (e.g., at 99.99th percentile).

Because the distributional approach is data- and labor- intensive, this type of analysis should be considered as a "refining analysis" conducted only after the screening assessment using point estimates shows an unacceptable risk. We have the software to conduct a distributional approach for dietary exposures to a single chemical, but an elaborate system would have to be set up to address aggregate and cumulative exposures.

How to deal with default scenarios: The default factors for dietary water and non-dietary indoor and outdoor exposures should be employed rarely, if ever. First, there is scant evidence to support a 10 percent default for water. With a few exceptions, pesticides are not found in ground or surface water at levels to support this contribution to risk. Secondly, those exceptions are generally well known and U.S. EPA has data upon which to base an exposure estimate. For indoor and outdoor non-dietary exposures, knowledge of the use patterns should allow use of surrogate data from pesticides with similar use patterns.

Data call-in for moving away from defaults: To some extent, this will depend upon interim policies and defaults, and their underlying rationale. Data needs will be chemical-specific. Data call-ins could be studies of residues in milk, transplacental transport and mechanistic studies, drinking water residue monitoring, and the like. If the interim policies lead to extremely conservative risk assessments, it would be up to the registrants to decide which areas they want to refine. Would there be any chemicals left to deal with in the second scenario? Our guess is that if the second round requires input from an extensive data call-in, the information may not be available before a majority of tolerances are reassessed. What may be left are the new active ingredients, and the requests for reassessment after the registrants generate studies to address specific defaults in the interim policies.

Then there is the consideration of cholinesterase (ChE) inhibition as an endpoint itself. Aggregate risk for OPs and carbamates pose a good example. Since it has been concluded that there is insufficient data for subgrouping OPs, it appears that all OPs will be treated as one group. The SAP approved U.S. EPA's policy on ChE inhibition, which would allow the use of plasma ChE inhibition as an endpoint. We expect that it could well be used as the endpoint for cumulative risk assessment. But is it an adverse effect? Are these chemicals additive, synergistic, or antagonistic? We simply do not know the answers. U.S. EPA must consider that if its decision results in wholesale cancellations, there would be a major impact on a multitude of pest management systems. This in turn would put more pressure on remaining alternatives and threaten the viability of integrated pest management systems. A program driven by dietary concerns alone may wipe out pesticides that pose only a theoretical risk to the public, forcing pesticide users to turn to alternatives that pose little dietary risk--but that also may threaten the environment.

Conclusion

U.S. EPA, DPR and regulatory authorities in other states face intense criticism from pesticide users and activist groups if FQPA does not fulfill their expectations. We must assure increased protection for infants and children while keeping agricultural goods plentiful and affordable. If U.S. EPA expects to achieve these goals, the Agency must fully involve the states in FQPA's implementation process. DPR already is working with registrants to look at critical uses, and U.S. EPA has benefitted from our accurate information on actual use. Any successful dialogue on science and policy depends not only upon scientific expertise, but experience with the issues at hand. California and other states have demonstrated that their scientific resources and field experiences are needed to make FQPA succeed.

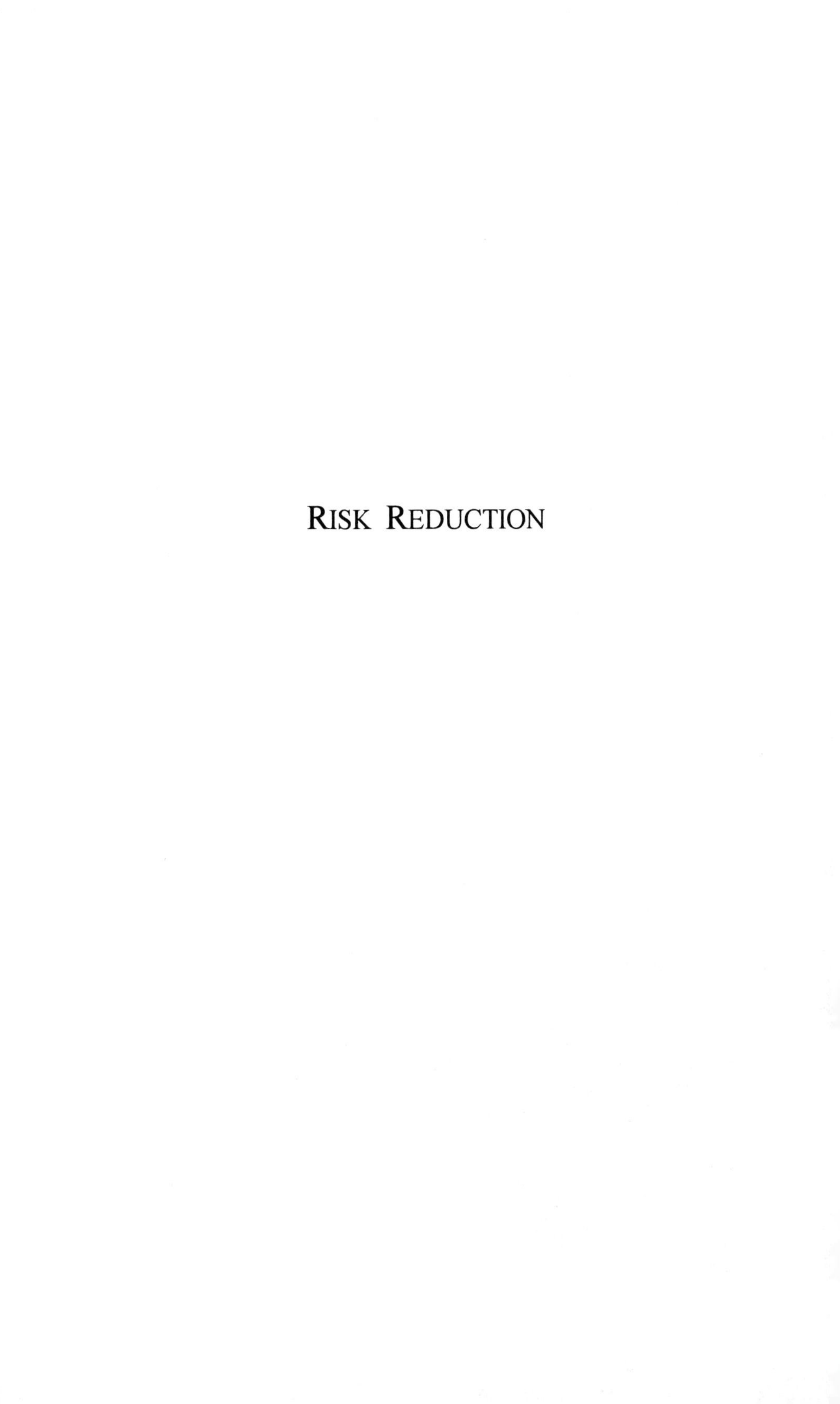

Risk Reduction

Chapter 7

Natural Products as Leads for New Pesticides with Reduced Risks

Gary D. Crouse

Dow AgroSciences LLC, 9330 Zionsville Road, Indianapolis, IN 46268

Although the use of naturally-derived secondary metabolites as pest control agents predates modern agriculture by centuries, technology-driven agricultural practices adapted during the past 50 years curtailed their use in favor of simpler, faster acting and more environmentally stable synthetics. Technological advances are again leading a trend back to naturally-derived materials as the industry learns how to find, optimize, and deliver naturally-derived pest control products. The result will be safer, less environmentally hazardous products for pest control.

It seems a natural fit—the best way to control pests is to learn from nature. Organisms are equipped with a broad array of chemical weapons to ward off predators and competitors for limited resources. The use of naturally-derived chemical pest control agents such as nicotine and pyrethrum clearly originated well over a century ago. A limited degree of success was derived from their use, and they were clearly preferable, from a toxicological standpoint, to many of the early chemical treatments based on toxic inorganics or petroleum distillates.

Nevertheless, widespread use of naturally occurring pesticides in modern agriculture has remained somewhat limited. As a class, biological control agents account for considerably less than 1% of the overall crop protection market (*1*). There are a number of good reasons for this. An organism that has no evolutionary pressure to confer such characteristics as photostability, mammalian selectivity or fit in to integrated pest management (IPM), did not build these attributes into its chemical arsenal. As a result, a plant-derived secondary metabolite, such as nicotine, shows little selectivity between insects and mammals. Even a product that is both safe and effective will often encounter difficulties in large scale agricultural applications. *Bacillus thuriengensis* (Bt), for example, is a family of highly selective proteinaceous

Adapted in large part from Crouse, Gary D., *CHEMTECH*, "Pesticide Leads from Nature," Volume 28, Number 11, November 1998, pp 36–45.

insect toxins, but the need for precise timing of applications has limited its effectiveness as a sprayable pest control agent (*2*).

It is difficult to argue that the origin of a material (synthetic or natural) can dictate the level of selectivity, compatibility with IPM, or any other measure of risk or benefit. Natural products, in fact, constitute some of the most toxic and most carcinogenic materials ever discovered (Table 1). Nevertheless, natural pesticides, in general, are perceived as being more environmentally suitable. Their superior environmental profile relative to synthetics is more due to historical differences in selection criteria. Synthetics are more residual because they were designed to be more residual, and they were used precisely because they could be applied across large areas and survive the effects of sun, air, water, and microbial degradation. In nature, the selection criterion is not durability, but simply whether a secondary metabolite confers a competitive advantage to an organism. In addition, greater structural complexity often results in both higher selectivity and greater fragility. Because biochemical pathways are so much more efficient than chemical synthesis, natural products have a tremendous advantage in the ability to produce highly complex molecules.

Table I. Mammalian and Environmental Hazards of Several Naturally-derived Pesticides

Botanical	Rat oral LD_{50}, mg/Kg	Other
Nicotine	50-60[a]	
Rotenone	10-350[a]	severe fish poison
Pyrethrum	1500[a]	severe fish poison
Physostigmine	4.5[b]	
Ryanodine	750[c]	highly toxic by i.p. injection
Abamectin	11[d]	moderate fish poison

SOURCES: [a] reference (*3*) [b] reference (*4*); [c] reference (*5*); [d] reference (*6*).

The real issue facing agricultural research has been one of how to adapt natural pest control methods to fit the needs of human agriculture. It is rare for a natural product to exhibit all the necessary characteristics for a commercially acceptable pest control agent. Success depends not only on their discovery but also on overcoming the technological hurdles of generating and delivering the activity where and when it is needed. The commercial and therapeutic value of penicillin, for example, was not able to be exploited until appropriate large scale fermentation technology was developed, some fifteen years after Fleming's initial discovery. Similarly, research into natural organisms has resulted in the discovery of myriad pest control solutions that have exquisite levels of efficacy, selectivity, and environmental safety. The technology necessary to bring many of these materials to the market has been successfully developed. Other opportunities remain tantalizing yet technologically unattainable even today.

The following review is organized according to technological limitations that have been successfully surmounted, those that have yet to be solved, and the roles these technologies play in expanding the role of naturally-derived pest control products. These technologies are divided into four subtopics:

- Finding useful biological activity from natural sources;
- Large-scale production technology
- Modifying natural materials to enhance activity or eliminate undesirable characteristics;
- Delivering the activity where it is needed.

Finding useful biological activity from natural sources

The ability to collect and screen a greater number and variety of organisms. It has been estimated that over 90% of the world's fungi and bacteria have not been taxonomically characterized, and many of the known organisms have not been adequately evaluated for potentially useful activity (*7*). Biodiversity screening began, logically, with higher plants. The successful development of human and animal health products from microbial sources led, in turn, to screening efforts against agricultural pests. Fungi, algae, sponges, insects, and higher animals have all produced a broad variety of activity and have recently received considerable attention as new sources of potential pharmaceuticals and agricultural leads (*8*).

More rapid identification and characterization of novel active entities. As a result of recent pharmaceutical and agricultural successes, the ever-widening natural product screening approach has flourished but produced critical bottlenecks in the lead identification process. The increased number of plants, bacteria and other organisms has brought the need for further miniaturization and more rapid assays. Within the pharmaceutical industry, first-pass large scale screening has largely been accomplished through the development of appropriate *in vitro* screens (*9*). The agricultural industry has been much less successful with this approach for a number of fundamental reasons. First, the physical demands on a pest control agent are much greater. Successful candidates are measured in terms of their ability not only to interfere with an important biochemical process but also to survive a battery of harsh environmental conditions. They must penetrate a plant, microorganism or insect and move within the organism to the molecular site of action. The difficult task of converting intrinsically potent molecules with weak whole organism activity into product-level agrochemicals has rarely been accomplished. A second disadvantage of *in vitro* screening of natural products in the agricultural arena is that it cannot detect new modes of action (MOA). Because of the problems of resistance, a new MOA is one of the most valuable attributes that a new agricultural entity can have. Finally, a key advantage of agricultural screening relative to the pharmaceutical screening is the ability to do without surrogates or model systems. With *in vitro* assays, that advantage is lost. Poor translation of *in vitro* activity within pharmaceutical and animal health screens is also observed regularly (*10*), however, the *in vivo* alternative is not available. Obligate parasites continue to present problems, because they still require surrogate pathogens or reliance on *in vitro* approach.

The development of better microanalytical techniques requires, in turn, the need for increased miniaturization of the screens. Successful miniaturizations of many whole plant, insect and fungal organisms into 96-well formats have been reported. With the increased ability to screen larger and larger libraries of organisms comes the technical difficulty of eliminating known active materials. Isolation and characterization of a molecule is often the most difficult part of the process, and even more critical when trying to characterize secondary metabolites that are present in very small quantity. This problem has prompted the development of the hyphenated-analytical techniques, such as LC-MS (liquid chromatography-mass spectrometry) and LC-NMR (*11*). Recent advances that include NMR solvent suppression and greater sensitivity have significantly enhanced the potential to rapidly identify known materials through routine processes. Coupling of LC and mass spectral fragmentation patterns with high speed computational capability allows for precise and rapid characterization of secondary metabolites (*12*).

From this expanded collection, screening, and characterization effort have come a tremendous variety of new active families of molecules. Many of these structures have been described in recent monographs (13, 14). Figure 1 lists examples of either new structural families or of older molecules for which pesticidal activity has been recently reported.

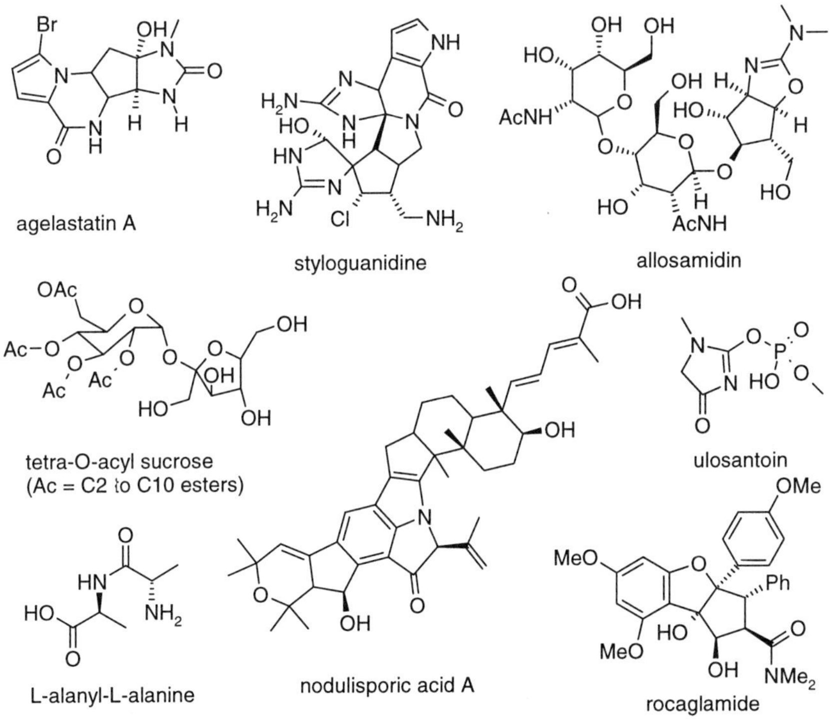

Figure 1. Recent examples of new natural products active against agricultural pests.

Agelastatin A, isolated from a marine sponge, has shown toxicity to members of the lepidopteran family (*15*). A member of the relatively large class of bromopyrroles, agelastatin A is the first to have been reported to be toxic to insects.

The phosphorylated hydantoin ulosantoin was also isolated from the marine sponge *Ulosa ruetzleri* (*16*). Styloguanidine, a potent chitinase inhibitor, was isolated from the marine sponge *Stylotella aurantium* (*17*). This alkaloid has not been screened for insecticidal activity, but another family of chitinase inhibitors, the allosamidins, show acaricidal and insecticidal activity (*18*). A new class of fermentation-derived indole terpenes, represented by the structure of nodulisporic acid A, have been reported to control blowfly and mosquito larvae at sub-ppm levels (*19*). Higher plants also continue to yield new structures. Rocaglamides, isolated from *Aglaia duperreana*, have been reported to have insecticidal activity equivalent to the azadirachtins (*20*). Structurally, however, they are much simpler than either the azadirachtins or ryanodines. Sucrose esters, found in leaves of the Solanacae family, have also been reported to have potent insecticidal activity. The number, length and position of acyl groups was found to be critical for whitefly activity (*21*). Pre-emergence activity has been reported for L-alanyl-L-alanine, one of a number of dipeptides found in corn gluten. Although not currently economically competitive with existing synthetic pre-emergence herbicides, this has been proposed as a natural alternative in locations where human contact cannot be avoided (22). All of these materials potentially represent some of the newer targets, and possibly new mechanisms, for chemical modification programs looking to develop new pest control agents.

Large-scale production technology

From initial screens to ultimate production, natural products present endless technical hurdles related to manufacturing. Fermentation screens usually are conducted on a milliliter scale, whereas as material requirements for a moderately successful agricultural product are likely to be in the hundreds or thousands of metric tons. Several fungicides, such as validamycin A and kasugamycin, are produced on a commercial scale in Japan. However, on a worldwide basis they are still considered relatively minor products.

Fermentation technology has progressed considerably in the last few decades, as demonstrated by the successful production of several agricultural products on a multi-ton scale (*23*). The microbial herbicide bialaphos (Basta®), is produced commercially by Meiji Seika through fermentation of *Streptomyces hygroscopicus* (*24*). This and several related di- and tripeptides are non-selective post-emergence herbicides with extremely low mammalian toxicity. The avermectins, also produced through fermentation, were developed initially as endectocides for animal and human health. Agricultural applications, as a bait formulation for control of the red imported fire ant (*Solenopsis invicta)* and as an agricultural miticide, were developed later. Although technical avermectin is itself quite toxic, its low use rate (15-30 g ai/ha) and rapid degradation in the environment result in a relatively safe product (*25*). Emamectin, a synthetic analog of avermectin, has a vastly different spectrum (Table II), and will soon be marketed for control of lepidopteran pests in high value markets. Mammalian toxicity is somewhat improved, while use rates are comparable to abamectin (*26*).

Table II. Activity of a Natural and Semi-synthetic Analog of Avermectin.

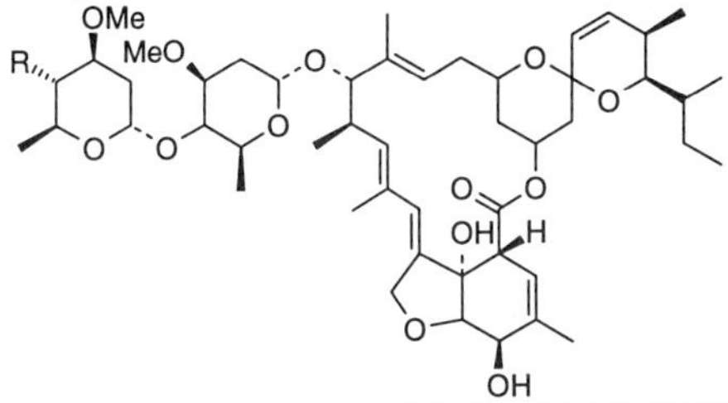

Derivative	R	TSSM Assay LC_{90} (ppm)	Southern Armyworm Assay LC_{90} (ppm) [a]	Rat acute oral LD_{50} (mg/Kg) [a]
abamectin	OH	0.03	8.00	11
emamectin	epi-$NHCH_3$	0.25	0.004	70[b]

SOURCE: [a] reference *25;* [b] reference *26*.

The spinosyns (Figure 2) are a class of highly selective fermentation-derived insecticidal macrolides, discovered in the early 1980's (*27*). The spinosyns are extremely effective at controlling members of the lepidopteran family of insects, which are major agronomic pests in such crops as cotton and vegetables. Just as importantly, all the other characteristics necessary for effective pest control, such as environmental compatibility, speed of action, low mammalian toxicity, and selectivity toward beneficial insects. (Table III).

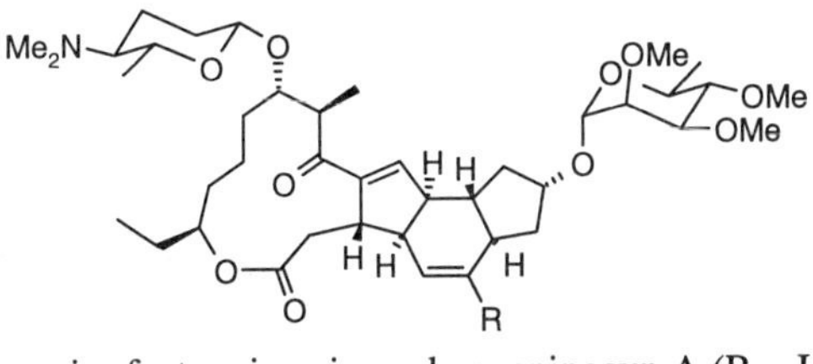

Figure 2. The two major factors in spinosad are spinosyn A (R = H) and spinosyn D (R = CH_3). To date, over 25 natural factors have been isolated and characterized.

Table III. Mammalian and Insect Toxicity of some Cotton Insecticides

	Rat Oral LD_{50} (mg/Kg)	Tobacco Budworm (TBW) LC_{50} (ug/g)
Spinosad	>2000[a]	1.12-2.4[a]
cypermethrin	250[b]	0.25-1.61[b]
Bacillus thuriengensis	>2000[c]	<1[c]

SOURCE: [a] reference *28*; [b] reference *29*; [c] reference *30*.

The organism that produces spinosyns is *Saccharopolyspora spinosa*, a slow-growing and aerobic gram-positive bacterium. These particular characteristics make large-scale fermentation more difficult than other fermentation media because of the potential for contamination by faster growing microorganisms. Nevertheless, a strain selection program has resulted in significant improvements in yield, resulting in successful multi-ton scale-up and commercialization in late 1997.

Modifying Active Natural Materials: "Designer" Natural Products.

Pyrethroids exemplified the primary drawbacks of natural product research: naturally-derived materials were not easily adapted to modern agriculture. Although known as safe and effective natural pest control agents, the pyrethrins were photolytically, oxidatively and hydrolytically unstable. Their relatively complex structure and fragility suggested unsuitability for agricultural applications. Furthermore, their complexity ruled out industrial-scale synthesis, thus the supply of material from plants was bound to remain a critical limitation.

A synthetic modification program was initiated by Elliot in Rothamsted (*31*) and by Sumitomo, and this effort eventually expanded to dozens of industrial and academic labs. The effort was eventually successful in developing an understanding of the structural requirements for activity, and synthetic analogs that overcame many of the drawbacks of the natural material (Table IV) were made. To date, over 35 synthetic pyrethroids have been developed for control of a wide range of insect pests (*32*).

The commercial success of the pyrethroids, in the early 1970's, came at a critical time for the agricultural industry. Traditional (random) chemical synthesis programs, responsible for a series of revolutionary pest control agents from DDT to organophosphates and carbamates, now experienced a plethora of new issues. Public concern about the health and environmental effects of highly residual synthetic pesticides was growing, as was the development of pest resistance to many of these same chemical classes. Biological control agents continued to play only a minor role, due to their high cost, low availability, and, ironically, too rapid environmental degradation. The pyrethroids demonstrated, for the first time, that the reduction of a complex natural product to a simpler, synthetically accessible molecule was possible. Not only were they simpler, but they were also more environmentally stable and up to 1000X more active than the natural pyrethrums, while retaining a reasonable mammalian safety profile.

The success of the synthetic pyrethroids prompted many research groups to set their sights on other classes of natural products. Today, these efforts establish natural product isolation and modification research as perhaps the most successful approach to the generation of new classes of pest control agents. The performance improvements relative to the natural product can be seen as fitting into one of two categories: improving the efficacy of intrinsically safe and selective natural products (Table IV), and improving the toxicity/environmental profile of effective but intrinsically unsafe materials (Table V). Research into synthetic juvenile hormone analogs has resulted in several commercial products. These demonstrate up to four orders of magnitude greater activity than the natural materials, as well as a broader spectrum and enhanced photostability. A potentially important new class of fungicides, based on the strobilurins, is now entering the marketplace. The natural fungal metabolite, strobilurin A, shows high greenhouse activity and excellent mammalian safety, but poor field efficacy due to high photolability. In a remarkably short amount of time,

Table IV. Examples of New Agricultural Pest Control Agents, and the Natural Materials on whose Activity they are Based.

Natural Product	Representative commercial synthetic analog	Characteristics optimized*
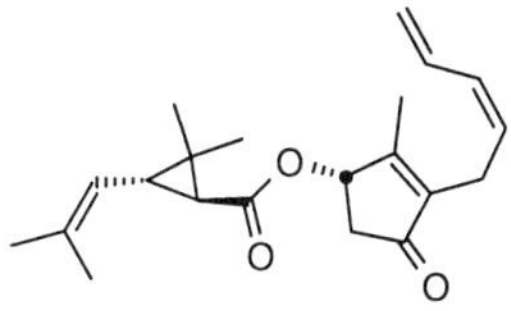 Pyrethrin I	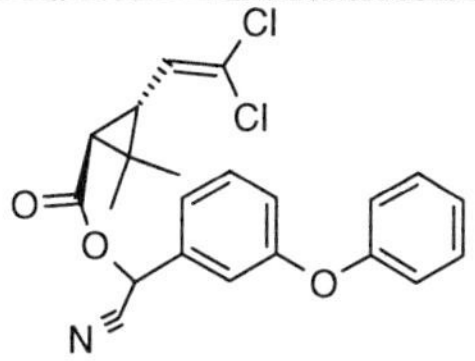 Cypermethrin	• environmental stability • pest spectrum • intrinsic activity (>1000X)[a]
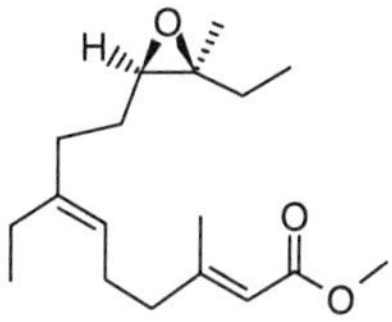 JH I	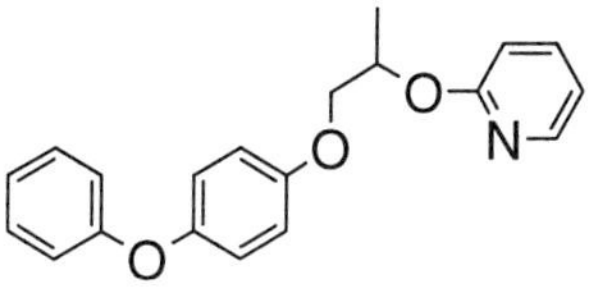 Pyriproxifen	• Photostability • Spectrum • Intrinsic activity (>1000X vs *aedes aegypti*)[b]
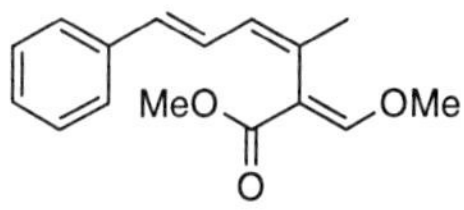 Strobilurin A	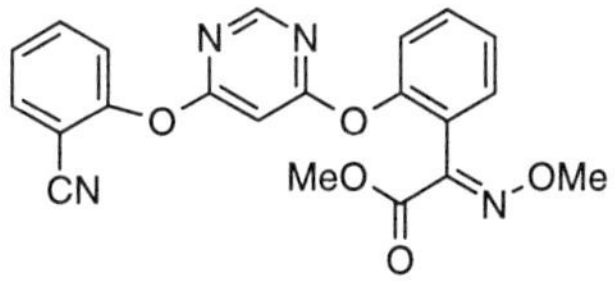 Azoxystrobin	• Photostability (ca. 7000X)[c] • Pest spectrum • Systemicity
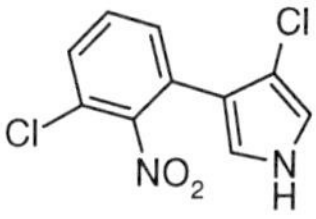 Pyrrolnitrin	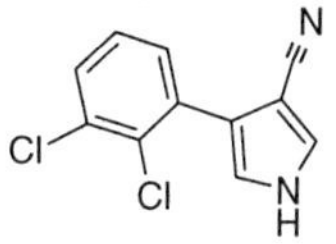 Fenpiclonil	• Photostability (100X)[d]
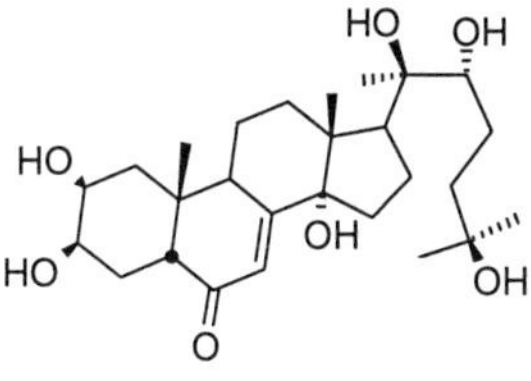 20-Hydroxyecdysone	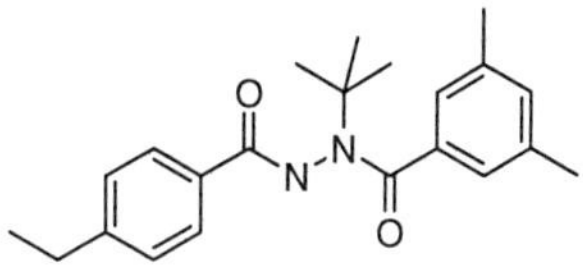 Tebufenozide	• Chemical simplicity • Improved transport • Metabolic stability (30-670X)[e]
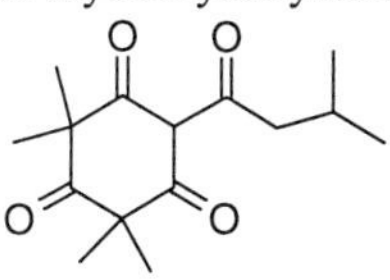 Leptospermone	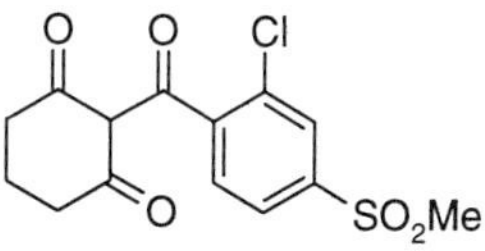 Sulcotrione	• Stability • Intrinsic activity[f]

* relative improvements are published measurements from laboratory comparisons.
SOURCES: [a] reference (*33*); [b] reference (*34*); [c] reference (*35*); [d] reference (*36*); [e] reference (*37*); [f] reference (*38*).

scientists at BASF, ICI, and elsewhere were able to develop a relatively simple pharmacophore model, and subsequently to prepare and evaluate numerous photo-stable bioisoteric analogs. The first members of the strobilurin class of fungicides entered the marketplace anly 13 years after the definitive structure elucidation of strobilurin A (*39*). Also, Pyrrolnitrin was the initial natural product from which two commercial fungicides fenpiclonil and fludioxonil are based (*36*). Leptospermone, a plant-derived essential oil, likewise provided the template for a series of p-hydroxyphenylpyruvate-inhibiting herbicides represented by sulcotrione (*38*).

The cases described above represent the commercial success stories from natural product modifications. Many other classes of commercial pest control agents are related to natural products, although more indirectly. The diacyl hydrazide tebufenozide, for example, was not discovered through a natural product structure modification effort, but was subsequently found to mimic the action of the natural ecdysone agonist 20-hydroxy-ecdysone (*37*).

Many other natural product modification efforts have been described which have not yet led to commercially successful products. A series of plant-derived lipophilic unsaturated amides that showed insecticidal activity and mammalian safety nearly equal to the pyrethrins was identified in the early 1970's. Despite a long and detailed synthesis program that succeeded in improving the environmental stability as well as the level of activity, commercially acceptable levels of control have not yet been achieved (*40*). Similarly, other natural products, including strigol (*41*), azadirachtin (*42*), hydantocidin (*43*), tenuazonic acid (*44*), ryanodine (*45*), and many others, have been the targets of numerous structural modification efforts without commercial success. Undoubtedly, continued efforts in one or more of these areas will eventually generate some of the next important classes of pest control agents.

The ability to successfully mimic proteins and polypeptides would create numerous opportunities for selective pest control. Many oligopeptides, proteins, and depsipeptides have been found that show selective control of insects, fungi, and plants. The pyrokinin and myosuppressin neuropeptide families (*46*), as well as insecticidal depsipeptide destruxins (*47*), the phytotoxic cyclopeptide tentoxin (*48*), and the insecticidal cyclopeptides vignatic acids A and B (*49*), have been recently reported (Figure 3). Delivery of peptide-based pest control agents is, in most cases, not possible using traditional spray applications. (As discussed below, they can be more effectively delivered through the use of genetically engineered plants). An alternative is to engineer a more appropriate synthetic analog that is more easily made and applied. Peptidomimetics, which are non-amide derivatives of amino acid-derived entities, have furnished numerous leads (*50*) and products (*51*) in the pharmaceutical arena. From an agricultural perspective, these approaches have not yet led to commercial successes. A synthesis and modeling study based on the herbicidal cyclic peptide tentoxin was initiated, however initial synthetic targets did not result in any biological activity (*52, 53*). Another report details an attempt to mimic the pharmacophore of the cyclodepsipeptide jaspamide (*54*). A synthetic analog, although active by injection, did not have any topical activity.

Whether the inherent structural complexity and additional stability and permeability factors will continue to limit the ability to mimic peptide-based derivatives is not clear. Combinatorial chemistry and parallel synthesis efforts are able

Table V. Examples of Natural Products and Synthetic Analogs with Improved Toxicological Profiles.

Natural Product or Lead Structure	Representative Synthetic Analog	Toxicity advantage*
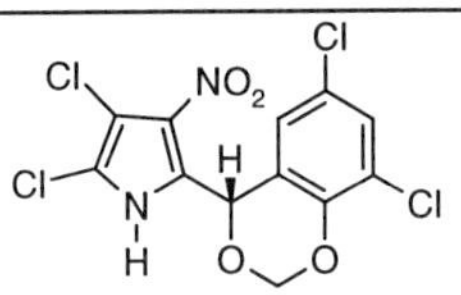 dioxapyrrolomycin	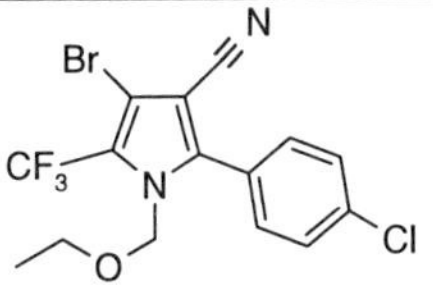 chlorfenapyr	mammalian toxicity (50X)[a]
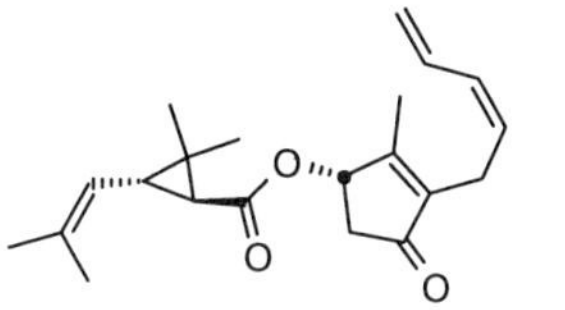 pyrethrin I	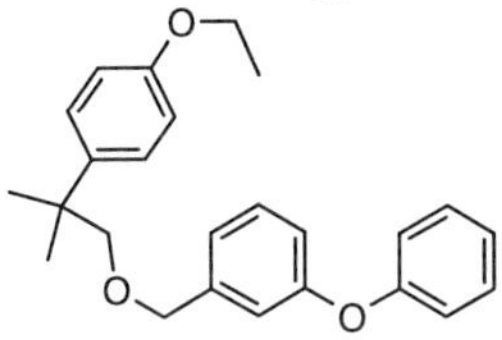 Etofenprox	fish toxicity (3000X)[b]
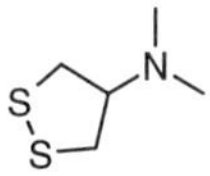 nereistoxin	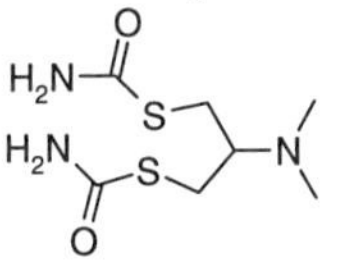 cartap	mammalian toxicity (2X)[c]
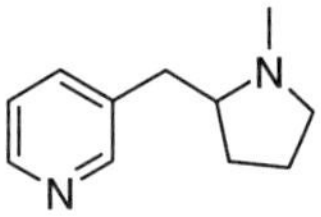 nicotine	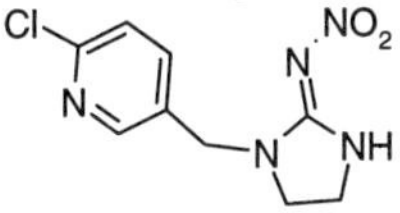 imidacloprid	mammalian toxicity (9X)[d]
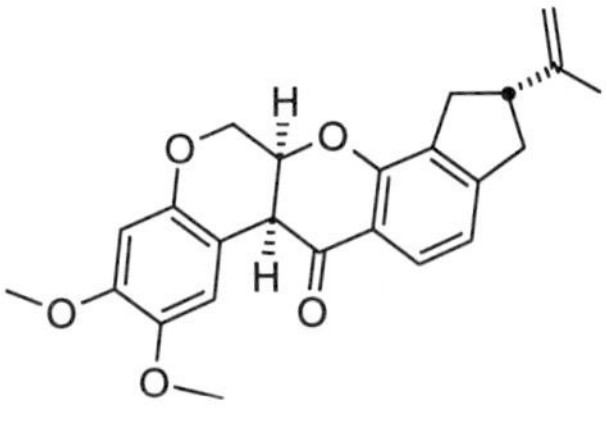 rotenone	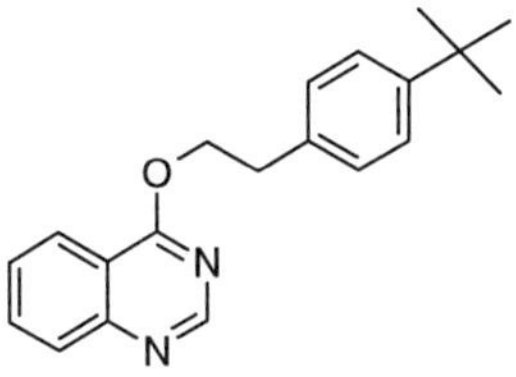 fenazaquin	fish toxicity (100X)[e]
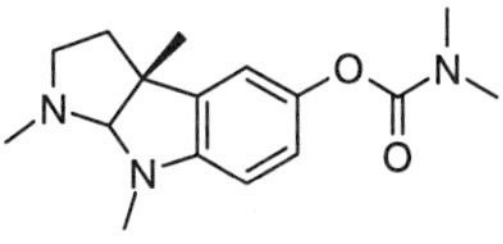 physostigmine	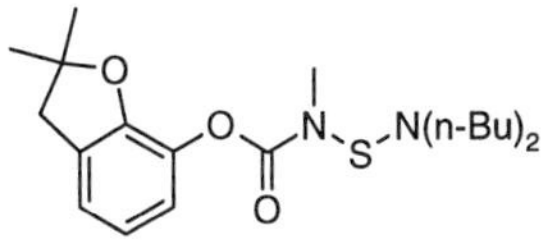 carbosulfan	mammalian toxicity (20X)[f]

*Toxicity advantage is defined as the ratio of rat (oral) or fish LD_{50} values for the indicated natural product relative to that of its associated synthetic analog.

SOURCES: [a] reference *(55)*; [b] reference *(56,57)*; [c] reference *(58)*; [d] reference *(59)*; [e] reference *(60)*; [f] reference *(61)*.

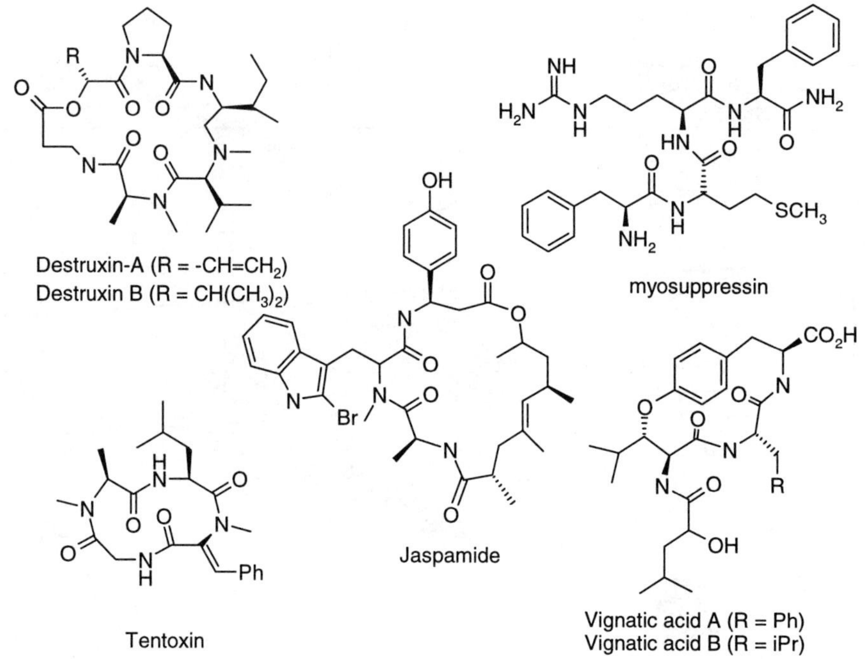

Destruxin-A (R = -CH=CH$_2$)
Destruxin B (R = CH(CH$_3$)$_2$)

myosuppressin

Jaspamide

Tentoxin

Vignatic acid A (R = Ph)
Vignatic acid B (R = iPr)

Figure 3. Cyclic peptides and depsipeptides with selective activity against insects or plants

to generate large numbers of peptoids and peptidomimetics, and expanded screening efforts will continue to evaluate these for biological activity. Clearly, much additional modeling work will need to be completed before design programs will be successful.

In the examples above, an intrinsically selective natural pest control agent was exploited by improving some physical attribute that limited its utility. Non-selective natural products have also be exploited by developing analogs that have greater selectivity. A key example again comes from pyrethroid research. Unacceptable fish toxicity associated with natural pyrethrins and many early pyrethroids limited their utility in aquatic environments. The non-ester pyrethroids, such as etofenprox, exhibit greatly reduced fish toxicity and can be used in aquatic environments (56, 57). Also, dioxapyrrolomycin was the starting point in research leading to the development of a safer and more active insecticide (*55*).

Table V also contains some commercially important pest control agents that, although subsequently found to share mechanistic and structural features with a natural product, were not derived directly from them. Imidacloprid, for example, may be considered an example of a natural product analog with reduced mammalian toxicity, even though the research leading to its discovery was not based on nicotine. Thus, even when natural products act at receptors which exist in both insects and mammals, analogs exhibiting adequate levels of intrinsic selectivity can sometimes be developed. Relative to rat muscle, insect nicotinic acetylcholine receptors are up to 1000X more sensitive to the effects of imidacloprid (*62*). Also, the several classes of miticides, including fenazaquin (Table V), exhibit target site binding characteristics similar to rotenone, even though structurally the two are quite different. Selectivity in this case is the result of differential metabolism (*60*). In retrospect, carbamates and organophosphates can also be formally considered as safer synthetic analogs of natural products, sharing some structural features and MOA with naturally occurring physostigmine and ulosantoin, respectively.

As in the discussion above, unacceptable toxicity can not always be overcome. In a paper relating to herbicidal sulfamoylated nucleosides, mammalian and plant activity were found to be too closely correlated (*63*). In general, natural pest control agents whose MOA involves disruption of primary metabolic processes for which there is a mammalian equivalent, are much less likely to lead to environmentally acceptable synthetic analogs.

Delivery of Activity

Genetic engineering: Beyond Bt. The inherent biodegradability of natural products, while considered an environmental advantage, creates a significant obstacle to delivering the activity where it is needed. Many natural products are simply too complex or fragile to ever successfully develop into a sprayable pest control agent. The ability to generate the active material within the crop or other organism solves numerous formulation and application problems. Following a lengthy development phase, recombinant technology has begun to generate dividends, with the successful introduction of pest resistant cotton and other crops containing the Bt gene. Their excellent efficacy and safety profiles (*64*) have contributed to rapid acceptance by growers and consumers.

A real advantage in genetic engineering thus lies in the value it can bring to other technologies. Further screening efforts into protein-based natural products can now be considered (*65*), not only as sources of leads for chemical modification, but as potential products themselves. Another strategy involving transgenic technology is the expression of plant genes for more complex plant defense materials. Lectins, which are primarily glycoproteins with specific carbohydrate-binding characteristics, have been shown to have insecticidal and nematocidal properties (*66*). Lectins active against *Homoptera*, *Diptera*, and *Lepidoptera* have been identified (*67*). Work is currently underway to clone and express these and other lectin-encoding genes in plants. Recent publications have indicated limited success with this approach, but the potential has been clearly demonstrated.

Other, even more complex, potential opportunities will become available as the tools for inserting and expressing genes becomes more routine. Plant-colonizing bacteria, initially considered as a potential delivery mechanism for Bt (*1*), have not yet resulted in commercial applications. Use of baculoviruses as vectors for delivery of an insecticidal gene is rapidly approaching the level of commercial applicability. Baculoviruses are insect-specific viruses that colonize and eventually kill their hosts. Their slow speed of kill has prevented widespread use for agricultural application. However, researchers have succeeded in incorporating a rapid acting spider venom into the baculovirus genome (*68,69*). Now, infection by the virus results in immediate expression of the toxin, leading to more rapid mortality. Alternatively, engineering viruses to express insect-specific regulatory hormones, such as, diuretic or anti-diuretic hormones and juvenile hormone esterase, have also been investigated with varying degrees of success (*70, 71*). The successful use of insect-colonizing fungal pathogens to control insects has also been demonstrated. The insect-specific fungus *Beauveria bassiana* was shown to control sweetpotato whitefly (*Bemisia tabaci*) in cotton (*72*). Genetic engineering to improve their efficacy has not yet been reported.

Ultimately, the incorporation into plants of genes which encode a series of proteins necessary for the production of secondary metabolites would appear to be an attractive, albeit more distant, approach. These technologies, if successful, will serve to blur even more the distinction between biological and chemical control methods.

Conclusion

Exquisite examples of selective toxicity can be found in nature. Organisms, with highly efficient means of producing complex chemistry and a few billion years in which to develop it, have developed an array of secondary metabolites that provide for themselves a competitive advantage. Humans have, in turn, taken advantage of many of these natural pest control entities for centuries, however many of the most selective and potentially most useful were too complex for commercial use. Technological developments in the last two decades have begun to unlock many of these heretofore unavailable natural biological control agents, and the next decade promises to expand considerably on this.

The ultimate effect that the recently passed Food Quality Protection Act will have on tolerances and uses of existing classes of pesticides is the subject of considerable discussion, some of which is presented elsewhere in this Symposium. It will certainly lead, in the short term, to some reductions in usage until appropriate interpretations and guidelines can be established. In the longer term, it is reasonable to anticipate that expectations for greater safety margins and lower residues will continue to increase, notwithstanding the significant gains already made. The need to develop products that that can be used in IPM programs will also increase the need for highly selective and readily degraded products.

Natural products do not necessarily represent an optimum in terms of efficacy, selectivity, or environmental suitability. Through advances in biotechnology, chemical design, fermentation technology and screening technology, naturally-derived pest control agents have been optimized for modern agricultural purposes, resulting in safer and more effective methods for controlling agricultural pests. The advent of newer delivery technologies, although mostly unproved, may provide an opportunity to capitalize on other highly selective but prohibitively complex plant protection agents.

Literature Cited

1. Fahey, J.W. In: *Biologically Active Natural Products; Potential Use in Agriculture*, Cutler, H.G., Ed., ACS, Washington, D.C., 1987, pp120-128.

2. Adams, L.F.; Liu, C.-L.; MacIntosh, S.C.; Starnes, R.L. In: *Crop Protection Agents from Nature. Natural Products and Analogues,* Copping, L.G., Ed., Royal Society of Chemistry, 1996, pp 360-388.
3. Matsumura, F. *Toxicology of Insecticides*, Plenum Press, New York, 1975.
4. Lynch, W.T.; Coon, J.M. *Toxicol. Appl. Pharmacol.*, **1972**, *21*, 153.
5. Dev, S.; Koul, O. *Insecticides of Natural Origin*, Harwood Academic Pub., Amsterdam, 1997.
6. Lasota, J.A.; Dybas, R.A. *Ann. Rev. Entomol.*, **1991**, *46*, 91.
7. Porter, N.; Fox, M. *Pest. Sci.* **1993**, *39*, 161.
8. El Sayed, K.A.; Dunbar, D.C.; Perry, T.L.; Wilkins, S.P.; Hamann, M.T., Greenplate, J.T.; Wideman, M.A. *J. Agric. Food Chem.*, **1997**, *45*, 2735.
9. *High Throughput Screening. The Discovery of Bioactive Substances,* Devlin, J.P., Ed., Marcel Dekker, Inc., New York, 1997.
10. Kirst, H.A. In *Progress in Medicinal Chemistry,* Choudhary, M.I., Ed., Harwood Academic Publishers, 1995, pp. 1-47.
11. Vogler, B.; Klaiber, I.; Roos, G.; Walter, C.U.; Hiller, W.; Sandor, P.; Kraus, W. *J. Nat. Prod.,* **1998**, *61*, 388.
12. Hostettmann, K.; Wolfender, J.-L. *Pest. Sci.*, **1997**, *51*, 471.
13. Hong, T.W.; Jimenez, D.R.; Molinski, T. F. *J. Nat. Prod.* **1998**, *61*, 158.
14. *Insecticides of Natural Origin,* Dev, S.; Koul, O., Ed., , Harwood Academic Pub., Amsterdam, 1997.
15. L.G. Copping, Ed., *Crop Protection Agents from Nature: Natural Products and Analogues*, Royal Society of Chemistry, Cambridge, 1996.
16. VanWagenen, B.C.; Larsen, R.; Cardellina, J.H., III; Randazzo, D.; Lidert, Z.C.; Swithenbank, C. *J. Org. Chem.*, **1993**, *58*, 335.
17. Kato, T.; Shizuri, Y.; Hitoshi, Y.; Endo, M. *Tetrahedron Lett.*, **1995**, *36*, 2133.
18. Blattner, R.; Gerard, P.J.; Spindler-Barth, M. *J. Pest. Sci.*, **1997**, *50*, 312.
19. Ondeyka, J.G.; Helms, G.L.; Hensens, O.D.; Goetz, M.A.; Zink, D.L.; Tsipouras, A.; Shoop, W.L.; Slayton, L.; Dombrowski, A.W.; Polishook, J.D.; Ostlind, D.A.; Tsou, N.N.; Ball, R.G.; Singh, S.B. *J. Am. Chem. Soc.*, **1997**, *119*, 8809.
20. Nugroho, B.W.; Gussregen, B.; Wray, V.; Witte, L.; Bringmann, G.; Proksch, P. *Phytochemistry*, **1997**, *45*, 1579.
21. Chortyk, O.T.; Kays, S.J.; Teng, Q. *J. Agric. Food Chem.,* **1997**, *45*, 270.
22. Unruh, J.B.; Christians, N. E.; Horner, H.T.; *Crop Sci.*, **1997**, *37*, 208.
23. Humphrey, A *Biotechnol. Prog.,* **1998**, *14*, 3.
24. Takebe, H.; Takane, N.; Hiruta, O.; Satoh, A.; Kataoka, H.; Tanaka, H. *J. Ferment. Bioeng.*, **1994**, *78*, 93.
25. Fisher, M.H. In *Phytochemicals for Pest Control,* Hedin, P.A., Ed., ACS, Washington, D.C., 1997, pp 220-238.
26. Technical Data Sheet, *Emamectin Benzoate Insecticide* 1995, Merck Research Laboratories, Agricultural Research & Development, Three Bridges, NJ 08887.
27. Kirst, H.A.; Michel, K.H.; Mynderse, J.S.; Chio, E.H.; Yao, R.C.;. Nakatsukasa, W.M.; Boeck, L.D.; Occlowitz, J.; Paschal, J.W.; Deeter J.B.; Thompson, G.D. In *Synthesis and Chemistry of Agrochemicals III*, Baker, D.R.; Fenyes J.G.; Steffens, J.J., Eds., ACS, Washington, D.C., 1992, pp. 214-225.
28. Dow AgroSciences Technical Guide, Dow AgroSciences, Indianapolis, IN, 1994.

29. Litchfield, M.H. In: *The Synthetic Pyrethroid Insecticides,* Leahy, J.P., Ed., Taylor and Francis, Philadelphia, PA, 1985, pp 99.
30. Adams, L.F.; Liu, C.-L.; MacIntosh, S.C.; Starnew, R.L. In *Crop Protection Agents from Nature. Natural Products and Analogues,* Copping, L.G., Ed., Royal Society of Chemistry, Cambridge, UK, 1996, pp 360-388.
31. Elliott, M.; Farnham, A.W.; Janes, N.F.; Needham, D.H.; Pulman, D. A. *Nature (London),* **1974**, *248*, 710.
32. Naumann, K. *Pest. Sci.*, **1998**, *52*, 3.
33. Elliott, M. In *Crop Protection Agents from Nature: Natural Products and Analogues*, Copping, L.G., Ed., Royal Society of Chemistry, 1996, pp.254-300.
34. Henrick, C.A., In *Agrochemicals from Natural Products*, Godfrey, C.R.A., Ed., Marcel Dekker, Inc., 1995, pp 147-160.
35. Sauter, H.; Ammermann, E.; Roehl, F. In *Crop Protection Agents from Nature. Natural Products and Analogues*, Copping, L.G., Ed., Royal Soc. of Chem., Cambridge, UK, 1996, pp.50-81.
36. Nyfeler, R.; Ackermann, P. In *Synthesis and Chemistry of Agrochemicals III*, Baker, D.R.; Fenyes, J.G., and Steffens, J.J., Eds., ACS, Washington, D.C., 1992, pp 395-404.
37. Wing, K.D.; Slawecki, R.A.; Carlson, G.R. *Science*, **1988**, *241*, 470.
38. Lee, D.L.; Prisbylla, M.P.; Cromartie, T.H.; Dagarin, D.P.; Howard, S.W.; Provan, W.M.; Ellis, M.K.; Fraser, T.; Mutter, L. C. *Weed Science, **1997**, 45*, 601.
39. Godwin, J.R.; Anthony, V.M.; Clough, J.M.; Godfrey, C.R.A. *Proceedings of the Brighton Crop Protection Conference—Pests and Diseases*, **1982**, *1*, 435.
40. Addor, R.W. In *Agrochemicals from Natural Products*, Godfrey, C.R.A., Ed., Marcel Dekker, Inc., 1995, pp 1-62.
41. Thuring, J.W.J.F.; Bitter, H.H.; de Kok, M.M.; Nefkens, G.H.L.; van Riel, A.M. D.A.; Zwanenburg, B. *J. Agric. Food Chem.*, **1997**, *45*, 2284.
42. Ley, S.V., In *Recent Advances in the Chemistry of Insect Control II,* Crombie, L., Ed., Royal Society of Chemistry, Cambridge, 1989, pp. 90-98.
43. Johnston, R.; Crouse, G.D. In: *Synthesis and Chemistry of Agrochemicals V*, Baker, D.R. and Fenyes, J.G., Eds., ACS, Washington, D. C, 1998, pp 120-133.
44. La Croix, E.A.S.; Mhasalkar, S.E.; Mamalis, P.; Harrington, F.P. *Pest. Sci.,* **1975**, *6*, 491.
45. Jefferies, P.R.; Yu, P.; Casida, J.E. *Pest. Sci.,* **1997**, *51*, 33.
46. Nachman, R.J.; Roberts, V.A.; Lange, A.B.; Orchard, I.; Holman, G.M.; Teal, P.E.A; In *Phytochemicals for Pest Control,* Hedin, P.A., Ed., ACS, Washington, D.C., 1997, pp 277-291.
47. Gupta, S.; Roberts, D.W.; Renwick, J.A.A. *J. Chem. Soc. Perkin Trans. 1*, **1989**, *12*, 2347.
48. Lax, A.R.; Shepherd, H.S.; Edwards, J.V. *Weed Technol.,* **1988**, *2*, 540.
49. Sugawara, F.; Ishimoto, M.; Ngo, L.-V.; Koshino, H.; Uzawa, J.; Yoshida, S.; Kitamura, K. *J. Agric. Food Chem.*, **1996**, *44*, 3360.
50. Adang, A.E.P.; Hermkens, P.H.H.; Linders, J.T.M.; Ottenheijm, H.C.J.; van Staveren, C.J. *Recl. Trav. Chim. Pays-Bas*, **1994**, *113*, 63.
51. Ohta, Y.; Shinkai, I. *Bioorg. Med. Chem.*, **1997**, *5*, 461.
52. Bland, J.M.; Edwards, J.V.; Eaton, S. R.; Lax, A.R. *Pest. Sci.,* **1993**, *39*, 331.

53. Cavelier, F.; Verducci, J.; Andre, F.; Haraux, F.; Sigalat, C.; Traris, M.; Vey, A. *Pest. Sci.*, **1998**, *52*, 81.
54. Kahn, M.; Nakanishi, T.S.; Lee, J.Y.H.; Johnson, M. E. *Int. J. Peptide Res.*, **1991**, *38*, 324.
55. Addor, W.R.; Babcock, T.J.; Black, B.C.; Brown, D.G.; Diehl, R.E.; Furch, J.A.; Kameswaren, V.; Kamhi, V.M.; Kremer, K.A.; Kuhn, D.G.; Lovell, J.B.; Lowen, G.T.; Miller, T.P.; Peevey, R.M.; Siddens, J.K.; Treacy, M.F.; Trotto, S.H.; Wright, D.P., Jr., In *Synthesis and Chemistry of Agrochemicals III,* Baker, D.R.; Fenyes, J.G., and Steffens, J.J., Eds., ACS., Washington, D.C., 1992, pp 283-297.
56. Udagawa, T.; Numata, S.; Oda, K.; Shiraishi, S.; Kodaka, K.; Nakatani, K. In *Recent Advances in the Chemistry of Insect Control*, Janes, N.F., Ed., 1985, pp. 192-204.
57. *Pyrethrum, The Natural Insecticide*, Casida, J. E., Ed., Academic Press, New York, 1973, pp 161.
58. Konishi, K. *Agr. Biol. Chem.*, **1970**, *34*, 926.
59. Elbert, A.; Overbeck, H.; Iwaya, K.; Tsuboi, S. *Proc. Br. Crop Protection Conf.—Pests and Diseases,* **1990**, *1*, pp 21.
60. Hackler, R.E.; Hatton, C.J.; Hertlein, M.B.; Johnson, P.L.; Owen, J.M.; Renga, J.M.; Sheets, J.J.; Sparks, T.C.; Suhr, R.G. In *Synthesis and Chemistry of Agrochemicals V,* Baker, D.R. and Fenyes, J.G., Eds., ACS., Washington, D.C, 1998, pp 147-156.
61. *Insecticides*, Hutson, D.H.; and Roberts, T.R., Eds., Wiley & Sons, Chichester, Great Britain, 1985; Volume 5.
62. Zwart, R.; Oortgiesen, M.; Vijverberg, H.P.M. *Pestic. Biochem. Physiol.*, **1993**, *48*, 202.
63. Kristinsson, K.; Nebel, K.; O'Sullivan, A.C.; Pachlatko, J. P.; Yamaguchi, Y. In *Synthesis and Chemistry of Agrochemicals IV,* Baker, D.R.; Fenyes, J.G., and Steffens, J. J., Eds., ACS, Washington, D.C., 1995, pp 206-219.
64. McClintock, J.T.; Schaffer, C.R.; Sjoblad, R.D. *Pest. Sci.*, **1995**, *45*, 95.
65. Gatehouse, A.M.R.; Gatehouse, J.A. *Pest. Sci.*, **1998**, *52*, 165.
66. Burrows, P.R.; Barker, A.D.P.; Newell, C.A.; Hamilton, W.D.O. *Pestic. Sci.,* **1998**, *52*, 176.
67. Zhu, K.; Huesing, J.E.; Shade, R.E.; Bressan, R.A.; Hasegawa, P.M.; Murdock, L.L. *Plant Physiol.*, **1996**, *110*, 195.
68. McCutcheon, B.F.; Hoover, K.; Priesler, H.K.; Betana, M.D.; Herrmann, R.; Robertson, J.L.; Hammock, B.D. *J. Econ. Entomol.,* **1997**, *90*, 1170.
69. Stewart, L.M.D.; Hirst, M.; Lopez-Ferber, M.; Merryweather, A.T.; Cayley, P.J.; Possee, R.D. *Nature (London)*, **1991**, *352*, 85.
70. Hammock, B.D.; Bonning, B.C.; Possee, R.D.; Hanzlik, T.N.; Maeda, S. *Nature (London)*, **1990**, *344*, 458.
71. Maeda, S. *Biochem. Biophys. Res. Commun.*, **1989**, *165*, 1177.
72. Wright, J.E.; Knauf, T.A. *Proc. Br. Crop Protection Conf.—Pests and Diseases,* **1994**, *2*, pp 45.

Chapter 8

Precision Farming: Technologies and Information as Risk-Reduction Tools

Franklin R. Hall

Laboratory for Pest Control Application Technology, The Ohio State University, Wooster, OH 44691

Precision farming (PF) or site-specific farming (SSF) is a relatively new concept in the management of production agriculture. Rooted in "management", PF brings the information age and space-age technology together with the science of producing food and fiber. In this process, there exists an increased opportunity to reduce environmental risks from pesticide use with improved environmental stewardship and greater economic profitability via more efficient use of scarce resources. PF thus is an example of combining newer techniques of satellites, remote sensing and computers with the familiar tools of soil testing, scouting and yield analyses. Information can help farmers reduce pesticide use, lower the need for insurance sprays, cut input costs and bring added environmental and economic soundness to the forefront. In order to exploit thc potential foı PF in accurately locating spatially variable pest/weed populations, a system of selectively applying cpas' (N, etc) is thus required. Development of "patch" sprayers, which are connected to computer-linked mapping, could allow treatment to patches within the field. Requirements for information acquisition, analysis, strategy development, delivery and evaluation of results can be very intensive depending upon the state of knowledge about the soil, N needs, as well as pest identification and development profiles. There remain serious questions about information ownership and economic benefits to all farmers. The various levels of information have to be well integrated in order to achieve an understanding about crop health and pest abundance/damage interactions, pest aggressiveness and invasion capacity, needed for a sustainable agriculture.

The increased world population in the next 25 years will necessitate increased food production approximating a 60% increase over current levels of production. Crop production chemicals have played a major role in achieving our current successes. However, critics of agrochemical strategies (1-3) have maintained that in spite of ca $25 billion for crop protection agents (cpa), which contribute to both human and environmental risks, crop losses continue at an alarming rate. Megatrends, such as capitalization, economics, technology transfer, policy, environmental and legislative issues, global competition, and emerging technologies will all significantly influence the potential for future cpa's as well as the structure of agriculture itself.

Risk reduction of our current crop protection tools and how well farmers control the use of these chemicals is governed in part by the regulatory structure as well as perception by the public. In some minds, **risk reduction is synonymous with reduced reliance on pesticides** (3-5). However, one of the prime benefits of cpa's is the reliability of consistent crop protection from year to year and thus reducing the uncertainties at the farm level. US production stability and price performance is dependant upon this functional precept and use of such technologies. This has led to the lowest consumer food prices in the world for the US.

Precision farming (PF) also called site-specific farming or prescription farming is an emerging technology of managing agricultural resources and production information. Concurrent with the hype brought on by the attractiveness of new "WOW" technologies is the implication that this new technology will immediately greatly reduce the imprecision currently undertaken by our farmers as they use and deliver agrochemicals. This brief review summarizes the technology and the key implications of the technologies, our current state of knowledge about the manageability of the technology and the anticipated effectiveness of economic and environmental gains from stepping into the next millenium with this technology.

Risk Reduction

Risk analysis of the hazards of new technologies has been placed into 3 phases including risk identification, risk estimation, and risk assessment. Flora (6) suggests that risk assessment criteria may be different for various disciplines. Additionally, costs and benefits of crop protection tactics are also different depending upon farmer risk aversions and his marketing goals, crop quality requirements etc.. Thus who pays and who benefits from crop protection are questions for sociologists, but disciplinary interactions are still lacking. Risk estimates of pesticide use are among the most controversial as the rules for health risks Vs that of environmental risks vary enormously among disciplines. Our ability to detect is far greater than our understanding of the risk itself.

EPA (7) presents an excellent review of the ecological risk assessment process. Ecotoxicology is receiving greater attention recently because of the off-target movement of these more active materials. While increased selectivity is the norm with these new pesticides, increased environmental safety is also a common benefit. Adequate quantification of the environmental effects of complex interactions of chemical combinations on a spatial/temporal scale is a very difficult process (8-9). Key questions on diversity, spatial and temporal sensitivities, recoverability of the systems (frequently ignored), and loss impacts of interacting pests require enormous data gathering and analyses especially when collected as various levels of "patches". Thus while the data is increasingly abundant, information and understanding for predictive tactical/strategic action is the weak link. Additionally, Marz (10) correctly identifies the significant differences and value of crop protection information from such sources as surveys, market surveys, station trials Vs on farm trials, on-farm information, and personal farmer interactions, etc. If risk reduction is based upon the premise of reduced usage of pesticides (rates, and frequency), then the value of such information in reducing uncertainties associated with reduced dependence upon cpa's is abundantly clear. PF is predicted to aid this more efficient use of pesticide/fertilizer tools of agriculture.

The current focus on input reductions and reduced reliance on single control strategies (directional policies, global economy pressures or regulatory issues) will require elaborate and comprehensive benefits assessments of any new technologies. Perceptions by the public about produce quality and federal/broker grading guidelines and use risk aversion, all impact growers' prices that, in turn, modify rates of technology adoption. Pesticides are relatively cheap insurance tools, fast and convenient technologies (11). The risks, particularly in fruit and vegetable crops, are high and pesticides save valuable management resource time (off -farm income requirements). Pesticide use strategies are influenced by public policies - commodity support programs, quality standards, global marketing, and disincentives for diversification, and finally, respond to price, although current prices do not reflect environmental costs (3,9). As summarized recently by Hall (12), pesticides are spectacularly effective and easy to use in order to respond to increasing food demands of an expanding population and reduce uncertainties in an already risky venture - agriculture.

Under risk reduction policy objectives, one could ask what is the risk, to whom and how do we define it (13)? If perception is reality, then the general public considers the use of pesticides as an insurance against crop loss as a serious risk and farmers use the technology too heavily. Solving the Delaney problem with the enactment of the Food Quality Protection ACT (FQPA) while addressing some problem areas has opened the door for many other questions. These include the serious implications of how to reduce risks from the elimination of organophosphates and carbamates with the few alternatives available to the fruit and vegetable and other minor crops (4,14). Pesticide reduction scenarios thus take on various roles of --- from what to what, and engender the needed questions --- *how, when, at what costs, to whom, and at what risk(s) to the farmer and food production industry*. Overuse of cpa's is not considered the problem by farmers faced with million dollar investments already vulnerable to such risks as meteorological variables (the 1997-1998 growing seasons). Thus, reductions in uncertainties, by using documented and proven technologies is not an unreasonable approach to improving food quality and production goals. PF seems to offer some potential to achieve these longer-range goals.

PF as an Emerging Technology

PF is a systems approach as illustrated succinctly by Parkin and Blackmore (15), and may require deeper understanding of processes to achieve a specific goal (Figure 1). Spatial variation of crops, pests, organic matter, yields etc within a field has long been recognized as a feature of farmland variables. Until recently, little has been done to exploit this variability by using improved management with emerging advances in technologies. Treating whole fields is an easy methodology, which is now expanded to tactics which recognize pest explosions within the field and treat as needed. These illustrations show that PF can link improved control of pesticide and N application with information management to improve delivery of a crop production tactic/chemical to specific sites within a field. In addition, the goal might not be to maximize yield, but rather to maximize economic advantages within particular environmental/economic constraints. The general elements of PF suggest an economic push and an environmental pull towards input reductions. Several conferences on agricultural PF have thus far been published including an international symposia (16), an agronomic conference (17), and a national science committee report (18) and a summary written by academia and promoted by industry (19). These represent the state of the art of PF both in the US and Europe with many more conferences already organized for 1999-2000. At this time, the research focuses on the ability to vary inputs of agricultural production such as seed, fertilizer, and pesticides as adjusted for crop yields, tillage, planting regimes, and soil characteristics based on scouting, monitoring, and harvesting. Thus PF can be viewed within a cycle of processes for various functions throughout the production phases of crops (Fig. 2).

Global Positioning System. PF requires a spatial positioning locator with a technology originally developed within the military and consists of using satellite signals to define positions on earth. An excellent practical review is given by Morgan and Ess (19) on the practical aspects of GPS/GIS parameters of PF. GPS can be used in two modes, a standard single receiver mode and a more accurate differential mode (dGPS). GPS is the cheapest and easiest to use but has reduced

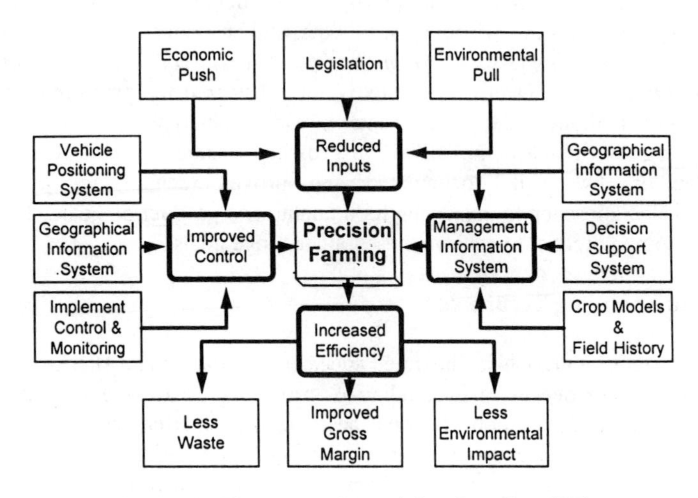

Figure 1. Elements of precision farming (PF)

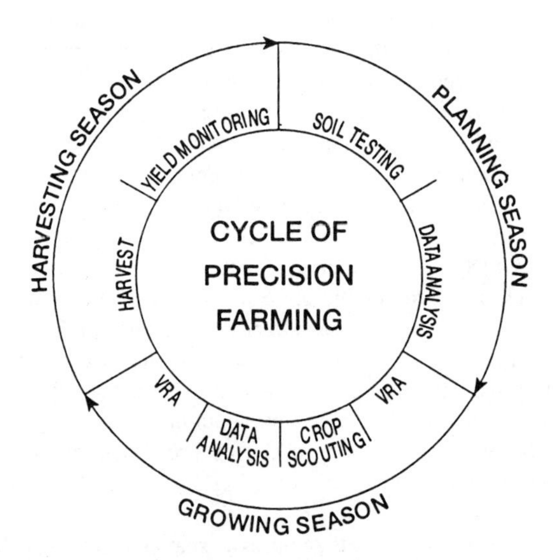

Figure 2. Cycle of processes in PF

accuracy due to positional errors (20 m vs 0.5 m). The dGPS has the second receiver located at a site different from the tractor mounted device. Current uses of the systems in agriculture mainly involve soil testing, field boundary mapping, positioning for soil types and fertilizer treatments, yield mapping and a more precise positioning of vehicles for applying fertilizers and pesticides.

Geographic information system (GIS). GIS integrates 2 types of data (1) spatial, which defines the shape and location of places, and (2) attribute data which describe organisms and things that happen there. Spatial data can be developed into vector data separating geographical features and rastor data to divide geographical units into cells and is used to map continuous data (Vs discrete). The various GIS software packages now on-stream such as ArcView, PS ARC/INFO, etc, are developing rapidly as is the ease of use and the power to seamlessly integrate various levels of information. Still weakly supported is how to use this information in predictive models and particularly for agriculture, the rapid (economic) collection of pest and weed data. New electronic sensors should provide some relief for this problem in the next 10 years. Easy to use decision-making analyses thus remains a key constraint to a more rapid acceptance of PF for agriculture

While a GPS can identify a location (of pests), it also needs a geographic information system (GIS) to tell is what that something is - i.e., patch of weeds, infestation, etc. --- if we have correctly identified and measured it. Storage of yield information, fertility levels, recommendation, etc. with GIS systems can provide it for every spot in the field. Thus a GIS is a software application that is designed to provide the tools to manipulate and display spatial data. GIS goes beyond just computerizing maps and can, with some use of overlays and linkages, combine data sets with agronomic models and decision support systems (Figure 3). Spot spraying and treat as needed tactics have long been a legitimate IPM strategies. This customizing of information about field health, productivity, etc, can now be managed to a much higher level. However, this integration of information is the current weak part of the system envisioned to reduce pesticide, etc. risks by identifying changes in management tactics.

Site-Specific Needs: Variable rate technologies (VRT), makes use of computer operated field equipment, which accurately delivers the correct amount of fertilizer/pesticide to a given point in a field. This increased flexibility in application equipment thus allows an infinite number of options to deliver a cpa on an "as needed" prescription-like basis. **Well-organized integrated information remains the key to optimizing this technology.** Coupled with new sensors infrared, optical, etc.), to identify targets within a field, an integrated delivery system with patch capability can selectively deliver appropriate concentrations of cpa's. This would increase precision of pesticide/fertilizer

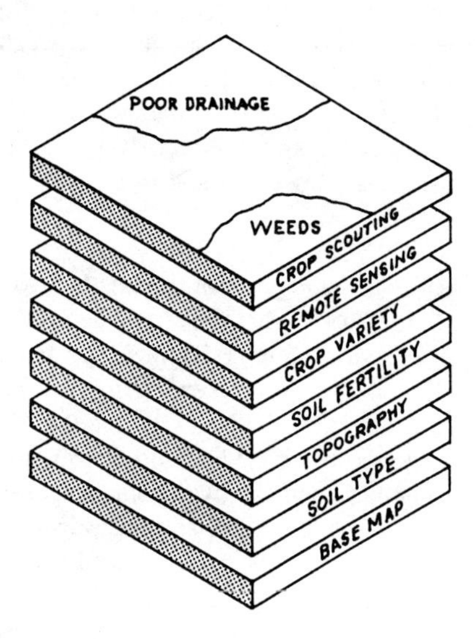

Figure 3. Data Layers in PF

placement both spatially, temporally, and in harmony with the environment, growth stage requirements and site-specific needs. Evaluation of these new innovations in sensors, VRT sprayers and GIS analyses are being summarized in reports of initial studies on PF (16-19).

Given the potential to have a wide range in pest management needs at various sites in a given field, then efficient delivery to that site is dependent upon the following advances in system interface between GPS and application (20).

Geoposition sensor: usually a dGPS system with up to .5 m accuracy such as the earlier Loral systems (21) and newer real time knematic systems (22) yield position accuracy's to ca 30 cm.

Ground Speed Sensor: usually by Doppler radar or wheel sensors allow accuracy's to 1-3% and support temporal resolution better than GPS and can take care of speed latency errors (Dickey John Co.).

Field Sensors: used by applicators to determine target rates, these include NIR organic matter sensors, and electro-magnetic induction sensors (16).

Target Rate-Field Log Maps: A number are now available for general GIS use and include vector (nets of polygon vertices, while raster formats have quicker data accesses. Management of regional data and transfer of data requires good communication/linkage packages with standards for rapid exchange since map accuracy, precision and legal issues are key concerns (23).

Applicator Rate Processing: rate commands are sent to the delivery system which include communication links for monitoring and control with appropriate standards ie; EIA-485, (20).

Networking: Various systems also include accurate distributed controls combined with standard interfaces of 3 pt hitches, and connections needed for any site specific delivery (24). New advances in fiber optics and wireless communications will add significant advantages to PF delivery systems (25). The limits of cost and field support are extremely relevant to the advancement and acceptance of PF technology throughout US agriculture. Neural nets and fuzzy logic control theory will add to this potential over the next few years (26).

Operator Interface: clearly an area which can be enhanced with visual cues at the seat to adjust for changes in the system. Virtual terminal access, etc will upon approval of standards and offer advances beyond the numeric/tonal displays now being offered. Again, costs are significant at this stage of development. However, the advances being shown to the public via the "Navigator" and "On-Star" systems being installed in personal cars at options costs of $1200 or so, offer hope that economical agricultural advances will follow.

Emerging technologies. Emerging spray application technologies (27) center around pesticide risk reduction thrusts as spray drift mitigation. These developments include air-assist nozzles, shrouds and adjuvants which increase drop sizes to reduce drift, as well as the new air induction nozzles and variable rate nozzles (Table 1). These advances in pesticide application technology (PAT)

Table l. Emerging spray application technologies

WHAT	HOW
◆ Drift Reduction	Air-assist, shrouds, low psi, adjuvants, EL, nozzles
◆ Control of Vol, Drop Size	Twin fluid, VR nozzles
◆ GPS/GIS	Monitor, site-specific, maps
◆ Pest Detection	Laser, etc. monitors
◆ VRT	On-the-go/site specific
◆ Decision Models	Trt mgt assistance
◆ Biotechnology	Plant resistance, value-added genome

Table 2. Expectations of pesticide application technology

◆ **Injector Nozzles**	⊕ Reduce drift, better performance, low cost
◆ **Air Assistance**	⊕ Reduce drift, better cover, perforance + savings
	⊖ High cost
◆ **Sensors**	⊕ Savings in AI, reduce drift
	⊖ High cost
◆ **Site-Specific**	⊕ Savings, as needed, records
	⊖ Still in development, cost benefits?
◆ **Model Processors**	⊕ Combine equip, low costs
	⊖ Costs of acquisition, limited
◆ **Electrostatics**	⊕ Savings?
	⊖ Robustness, flexibility, costs
◆ **Recycling**	⊕ Product savings, reduced ground + drift contamination
	⊖ Costs, needs dwarf trees
◆ **Inspection/Standards**	⊕ Savings + accuracy increases
	⊖ Costs to government / farmer?

have significant expectations (Table 2) in terms of the increases in delivery efficiency (hence reduced on-site and off-site risks) (28). Costs, legislation incentives and or public pressures will dictate whether these systems will advance into the practical realm of usage by the farmers of the next millenium. Improvements beyond the technical equipment parameters involve improved education and training of applicators for standard and new use procedures, cleaning and calibration practices, user protection and general Best Management Practices (BMP). Typically, the misapplication factors involve poor calibration, equipment malfunctions, and mismatch of chemical mixes with plant needs, etc. In reality, wrong field applications and spray drift account for 41% of all misapplications (29). **This suggests that PF technologies coupled with educational thrusts/farmer mentoring, etc, can improve the efficiency of the current cpa delivery process.** Although traditional leadership in these areas has been accomplished via government organizations, there is more pressure for the chemical industry to actively participate in these activities. Risk reduction issues (brought on by environmental and human exposure risk issues) focus on the modification of field edge practices, and the proper matching of equipment and plant needs with greater attention to an understanding of the potential technology solutions. Farmers are under serious economic pressure to improve the profit margins, ie., the 1998 grain and pig prices. Some government/insurance programs aimed at increasing farmer knowledge about crop protection options and adoption of risk reduction technologies, such as PF, could improve the rate of adoption where environmental issues are a priority (28).

Advances in Variable Rate Sprayers

The capability to define a specific location within a field, which may require additional fertilizer, seed, or pesticide treatment, then **requires the application system to deliver upon demand.** Poorly calibrated sprayers and variation in travel speed have resulted in numerous advances to control delivery, automation or controls, etc. in order to account for variables of speed, or changes in the rate of application. Anderson and Humbug (20) suggest 5 systems comprise an array of pressure-based flow controls, and various chemical injection and direct nozzle combinations. Both pressure and control based flows controls regulate the delivery of a chemical-carrier premix through the nozzles but which change the distribution across the boom swath if the nozzles are worn or damaged. An alternative is thus chemical injection where there is no premix to dispose of and multiple pumps handle the delivery with positive displacement pumps to meter agents into the carrier stream matched to boom width, travel speed, and desired broadcast rate. The principle limitation here is the transport delivery between injection point and nozzle discharge. Alternatively, placement of chemical injection points close to the boom and nozzles significantly reduce transport delays and volume changes (29), but which requires special non-traditional pipe networks and connections. Control systems can be developed to anticipate rate changes/delays and/or make changes early but requires spatial/temporal/logistic solutions. Direct nozzle injections appears to solve the delay dilemma (30) etc but require numerous additional chemical lines, and metering mechanisms, and a solution as to how to deal with the left-over material in these lines in disposal and clean-up etc (31).

Controlling both carrier and chemical has advantages in that response times are essentially the same as the response time of the flow control system. Currently, nozzles represent a limitation to this approach since both drop size and flow rate are functions of the pressure drop across the nozzle orifice. This creates a limited range of rates since pressures must be increased by 4 to achieve a doubling of the chemical application rate. Since spray drift with increasingly powerful actives is becoming a disturbing problem with urban growth, people and sensitive sites, (water, etc,), but this pressure range is unacceptable. New nozzle advances have shown the potential to overcome these serious problems in 2 separate ways; (1) adequate performance of nozzle over extended pressure ranges , as with the new designs – air induction low volume nozzles and /or, (2) an alternative solution shown by the Synchro system (33) which under constant pressure/volume can produce significant changes in the spray cloud characteristics, thus adding considerable flexibility (and toxin use efficiency) to crop/pest needs. The recent series of mergers of large equipment companies yield an uncertainty about the evolution of GPS/equipment technologies made available to certain commodity producers.

Information Requirements

Weeds represent a major target for patch spraying since weeds can occur in the absence of crop, in known patches, between crop rows and interspersed with the crop. Scattered about the field which can be followed with the use of weed infestation maps (labor intensive), or in the case of planted fields, improved sensors can identify any green thing as a weed, or newer sensors capable of differentiating weed structure is also possible. Such systems have shown remarkable reductions in pesticide use rates (90%), with even dual boom systems using 2 rate concentrations. More difficult than the between row sensing is the weed interspersed with the crop. Methodologies using reflectance and morphology have been successfully utilized but commercial systems are some years away (33). Perhaps even more intriguing is the discussion brought forth by Cardina et al (34) who ask for deeper research on "what if" advancing weed patches are broken up? Does this weaken competitiveness or strengthen weed advancement profiles? Thus short term and long term implications need to be considered on a total farm management basis (Fig. 6). Clearly, the information complexities brought forth by GPS/GIS capabilities will escalate enormously as the technology infrastructure strengthens.

Clearly information intensive, the acquisition and integration of the layers of widely different kinds of information (Fig. 3) into a practical usable model will be the acid test. Some critics of the technology still insist that this high-tech approach to crop protection is not the route to a sustainable, economically viable, and ecological sound agriculture. Many organizations and agencies i.e., US Geological Survey, USDA-ARS, and NRCS, US EPA, the University systems, and the crop protection and equipment industries with their dealer systems are waiting for the appropriate time to deliver this technology in the new millennium. While it is not necessarily prudent to be the first on the block, clearly there are signals about agriculture, information content and structure. Farmers, organizations, suppliers, etc, need to be both cognizant of and more importantly, prepared to initiate a step up in strategy assessments. **Historical trends, data, assessments, i.e., knowing where you are, where are the weakness on the farm (field/crop wise), and the what-if's of changing practices are all vital tools for impact assessments for new strategies**. Farmers need to be aware of data needs, the clarity of information, and establish clear "goal agendas" if they are to make effective use of these emerging technologies.

Biological data from GPS/GIS data acquisition can provide valuable clues about ecosystem viability and functionality, and the movement in space and time of pests/weeds. With the newer geostatistical tools, PF can help us integrate what is happening with proposed changes in crop protection strategies (34,43). A recent summary of GIS utilization case studies for managing natural resource landscapes (44) shows how local, state, national agencies as well as private industry are developing the framework for measuring and analyzing resource utilization. Complete with demonstration copies of ArcView GIS samples, sources of national resource web sites add to the power of well-organized and time forecasting. The current weakness in landscape spatial issues, is the point-data analyses which begin the more difficult tasks associated with theory, ie., identifying the scale and understanding the nature of the spatial structure. Pesticide resistance management and good stewardship demand a tracking of refuge and genetic engineered planting spatial relationships. Herein lies the additional worth of this technology as the influence of biotechnology on agriculture escalates with large-scale plantings of genetically-modified plants.

Remaining Questions About PF Technologies

The fragility of global financial markets in 1998 has shown remarkable direct influences on the buying strategies and hence, US grain prices. The weather (El Nino/La Nina) influences continue to demonstrate enormous power to disrupt vast US grain/food production potentials. Will technology be able to rescue the farmers from these and other factors influencing their profit margins in coming years? Alternatively, for some farmers, input reductions may be the only recourse of action. It remains an interesting "tug of tactics" between using technology adoption to provide solutions Vs input reductions via alternative agriculture tactics. Meanwhile there is continued pressure to grow more food for a burgeoning world population.

What are the remaining key issues to be resolved? They include: (1) identification of the economic benefits of spatially variable crop protection measures for the average farmer, (2) development of intelligent data-based models and simplicity of the analytical/prediction GIS assessments, (3) questions about yield mapping benefits for the majority of farms, (4) yield interpretation models, (5) the need for simplicity in managing spatially variable inputs to improve the measure of uncertainty values in data sets, and (6), the questions of farm size on technology adoption (35-37). The effort to organize farm records, with spatial functions to create a GIS is a useful exercise even if records stay in the shoe box and spatial data remains penciled in on aerial maps. It does stimulate thinking about farm goals, and deeper issues of why certain field areas contributed less than optimum to farm income. How well the GIS is organized, in the layering to accomplish this increased learning capacity for the managers, remains the weak point and the focus of many future symposia about PF. Data is abundant, but information remains obscure in that we require alternative statistical analyses even beyond geostatistical programs (35-36). Can we fine-tune farm inputs recognizing that economic analyses are incomplete so that beyond pesticides, genetic engineering, crop rotations, and diversity will achieve optimal yields without over-dependence upon "insurance" methodologies of pesticides?

Since the bulk of PF development is still coming from the private sector, **data privacy is a serious hurdle**. This is especially relevant when it becomes obvious that the data, which has value beyond the farm gate, is out of control by the individual farmer. The transferability of this information could have significant value to potential users such as other farmers, agribusiness, processing firms for new market trends, as well as extension education and regulatory oversight and could be a sensitive area. This is similar to events now occurring at the neighborhood grocery store which issues "advantage" cards for purposes of obtaining added discounts. The card also records user purchases, timing, and an array of information about food purchase decision-making. Finally, with biotechnology companies increasing their control of seed companies, this integration of US agriculture is likely to have a significant impact and change on traditional farmer crop production decisions.

As farm supports change (FAIR) and increase crop diversity, advisors become more valuable assets for planning farm tactics. Farmer goal setting processes thus have a unique opportunity to be reassessed by industries attempting to influence how, what and when they undertake a management practice. Critics of technology say it is much too narrow and we should be doing more studies of farmer behavior and goal setting components (38) especially under risk taking tactics. There still remains a misperception by the public of why pesticides are used in crop protection. This only exacerbates the on-farm dilemma of how far to take on new expenditures for risky production tactics under the current commodity prices.

Adoption rates of new technologies will impact on future developments of this technology. As illustrated by Lowenberg-DeBoer (39) in a historical perspective of adoption of corn hybrids, the adoption rate of PF is likely to have similar delays and problems unless we understand what influences technology adoption (Figs 4 and 5). Hewitt and Smith (35) summarize the concerns of many about PF trends, which on the surface represent a new way of managing ag resources and production information for the potential good of the environment and production goals. PF has certain characteristics which make it unique from other agricultural innovations. To date, PF has been primarily a private sector endeavor, and certain elements of the technology may never be controlled/owned by the producers. Effective on-farm use of the PF data continues to be extremely information-intensive (requiring a sophisticated manager, trained dealer) and this information has value beyond the farm (35). Can PF be made more scale neutral, aid input reductions (hence reduce environmental intrusions), and do more than just increase efficiency? Since information has power, can producers control their own destiny and management goals with the use of PF? Will PF exacerbate environmental and social costs associated with input-intensive agriculture (36)? This array of interesting questions about goal-setting, visionary farmers and the precision level of the average farmer, and potential for problem solving (35-37) may all be hidden in the reality of the global economic crisis, hence effects on US grain prices. These, among many questions, are typical of those being addressed in forums of International meetings (16-19), various reports, and News articles (40). They represent difficult, but practical questions which need to be addressed. As cries for increased effort from the public sector are raised (40), efforts by the Deere group and others (19), EPA (for N impacts), NRCS (soil conservation/erosion), and the USDA (production), are moving to support PF research needs, (37-39, 40). However, the jury is still out on whether, in fact, we can detect economic benefits from the array of PF management changes taking place within a field.

The National Research Council (18) and others suggest we need to focus on several goals. These multidisciplinary approaches to PF include: (1) **create data gathering and analysis tools** for agriculture, (2) **clarify intellectual property and privacy rights** with public organizations involved here, (3) **link rural connections** in the internet, (4) provide unbiased **assessments of economic and environmental impacts** of such sustainable management, and (5) **educate and train professionals.** Nevertheless, the technology is advancing rapidly to increase the precision of farmers and their interactions with mentors. These tactics are all aimed at "an improved use of information", both existing and layered. How well we organize that data, and prepare for active discussion on long-range assessments of technology, will insure PF a significant position in future US agricultural practices.

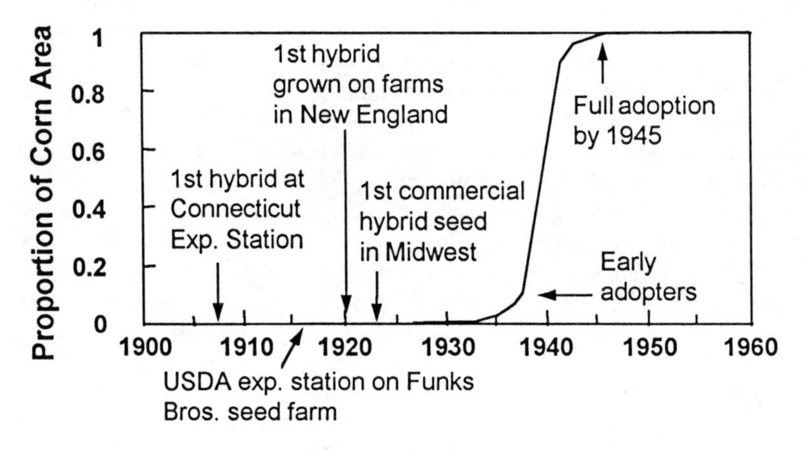

Figure 4. Adoption of hybrid corn in the US

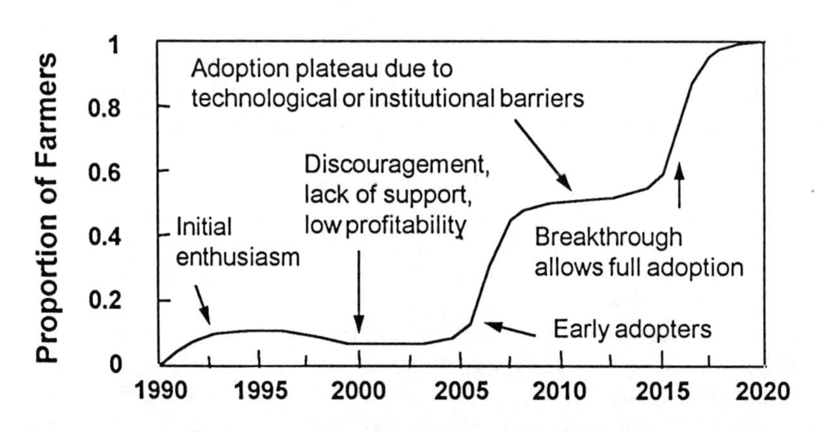

Figure 5. Predicted adoption of PF in the US

Dissemination of PF technologies will be aided by electronic information sites on the WEB. Selected resources include the mega links such as the ESRI web site at esri.com/arde which serves as a universal spatial server for ARC/INFO. ARC NEWS (ESRI) offers clear updated information on related sites, GIS news and insight into future offerings and developments about databases, new models, GPS interactions, etc. Other PF sites include: http:www.agriculture.com and http://nespal.cpes.peachnet.edu/pf/ . PF equipment systems resources, such as John Deere, Case, New Holland, AcChem, etc.,with the GPS dGPS sites of Rockwell, Trimble, Starlink, etc., and the GIS mapping dealers, MapInfo, Farm Works, ESRI, GIS world, etc. are the major players attempting to exploit PF for agriculture. VRT technologies offered by AgChem, Tyler, etc, and an excellent glossary provided in an easy to read guide on PF for farmers (19) all provide valuable resources. There also continues to be an escalation of world conferences hosted by the major scientific societies.

Beyond Pesticides/Fertilizer Placement Precision

The automatic sensing of pests and weeds among crops is a next logical development and require a "fast response" characteristic linked into delivery systems. Even greater benefits could come from positioning sprays within canopies in a form of "micro-targeting'. These systems could really lead to very low dose applications and a truly target –oriented approach to pesticide delivery (15) and significantly reduce AI/acre. The use of GIS technologies, however, offer potentials to go far beyond the development of more precise pesticide delivery systems for agriculture. These interesting uses range from the very practical to ecologically based hypothesis testing. **First**, there are the aquatic risk assessments for EPA, which conventionally utilize conservative worst case assumptions using the following parameter examples:

1. 10 ha watershed is 100% cropped with cotton,
2. Cropping areas occur up to the water edge
3. Maximum #applications at maximum rates applied by air
4. Winds blow towards the water and there is no marginal vegetation.
5.

Under the auspices of the pyrethroid working group and the exposure modeling group efforts, Zeneca and others in a team work environment (41-42) have attempted to use GIS to examine the landscape via satellite imaging to show:

1. size classification of individual fields,
2. margin relationship to water and crop
3. spatial distributions around ponds with directional components
4. buffer composition

These initial spatial and temporal risk assessment analyses show that (1) EPA's default landscape assumptions overestimate exposure, (2) many fields are never affected by the insecticide, and (3) **physical** buffers do exist and can mitigate drift and run-off. Thus, GIS tools have interesting value in (1) **determining how potential environmental exposures occur**, (2) can **characterize the agricultural landscape**, (3) can **provide verifiable data** for refining model assumptions and (4) **lend clues for the development of practical farm management action** plans using buffers, BMP and drift mitigation tactics. A higher level of analysis, however, will require increased integration of spatial crop information, crop production practices and environmentally realistic exposure predictions.

The **second** example of potential GIS usage involves the tracking of transgenic fields, wherein the spatial relationship of these plantings can be tracked relative to spatial orientations of non-transgenic field plantings (for resistance management, etc). Given the state of the number of acres now being planted to genetically modified seed, this would seem to be a logical research focus for the next 5 years. **Third**, ecological systems are spatially heterogeneous, with complexity and variability in time and space. A comprehensive review of maps and spatial point data analyses and interpretations strongly suggests there are opportunities to improve compatibility's of point-data analyses with ecological theory (43) including geostatistic alternatives to patch-based approaches when system properties vary with space and time. Using heterogeneity indices to advance the linkage between patterns and processes (scale of patchiness with geostatistic/fractal techniques) can add to advancing ecological theory. Cardina et al (34) also suggest the need for a greater understanding of the more intensive landscape management potentials of GIS technologies wherein patches of weeds are broken up thus creating a myriad of weed aggressiveness. Monitoring and managing decisions for long term views to minimize land use risks thus become much more information intensive and require increased research team communications beyond our current levels (Figure 6).

Summary

The ability to vary inputs in a field defined by a site-specific need is critical for any development of site-specific management. The technology to utilize variable rates is available. Soil maps showing soil types, integrated with N requirements, plus/minus historical crop responses, is a laudable goal in optimizing N utilization. Adding risk management tactics can be as easy as changing spraying tactics in high risk farm border areas to advancing weed patch tactics with as needed herbicides, to better understanding ecological dynamics of various pest, weed infestation movements. With VRT and economic/ecological/environmental impact assessments, there is an enormous jump in requirements of logistics, equipment, labor and management resources especially decision-making. This will require government support if projected advances in IT, PAT, GPS/GIS are to increase more rapidly than projected. While promising to reduce cpa usage, the potential increase in risks of crop failure may not be reduced unless we have greater

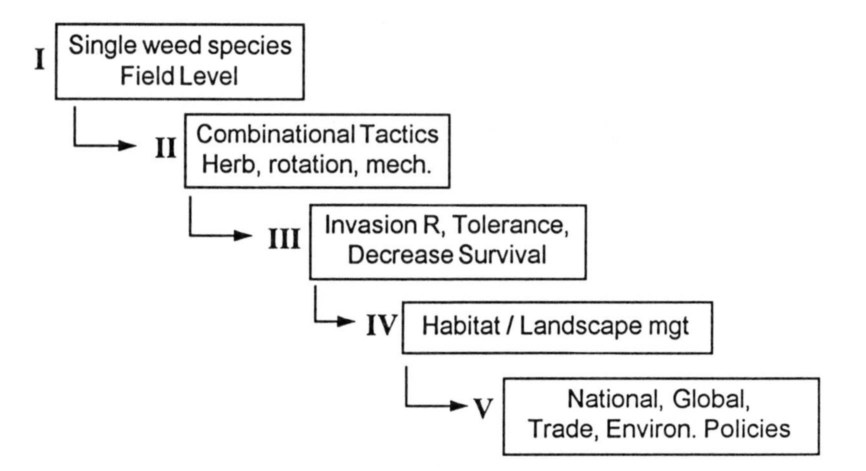

Figure 6. Advancing weed IPM strategies

attention to a wider arena of team research involving more stochastic processes. This problem will need to be addressed if such strategic goals are to be achieved.

Clearly, **economic benefits have thus far been a serious weakness** and as global economies and weather affect farm prices (as in 1998), then acceptance of these advances will continue to lag behind development. **Education, training, and the sensitive issues of data ownership and data privacy have to be addressed by public organizations**. IT via the WEB, new partnerships between agricultural cooperatives, or value-added resellers will add significant pressure to acquire this technology. Can we project that the grower, faced with increasing cost pressures, stay in business, organize a good future for his farm with reduced inputs or will it take extremely far sighted mangers to see the value of complexity made simple via IT with elegant GIS maps/projections. Can we also utilize PF to enhance IPM stage II strategies wherein field borders (invasion points) can be identified and treated differently than the inner sections (reduced pest infestations)? Clearly, the use of GIS for fundamental ecological/pest movement dynamics and risk reduction (chemical exposure scenarios) will enhance probabilities of acceptance. We shall see.

Literature Cited

1. Pimentel, D.; McLaughlin, L.; Zepp, A.; Lakitan, B.; Kraus, T.; Kleinman,P.; Vancini, F.; Roach, W.; Graap, E.; Keeton, W.; Selig, G. *Bioscience*, **1991**, 41, 402.
2. Pimnentel, D.; Acquay, H.; Biltonen, M.; Rice P., Silca, M.; Nelson, J.; Lipner,V.; Giordano, S.; Horowitz, A.; D'Amore, M. *Bioscience,* **1992**, 10, 750.
3. Pimentel, D. ed, *Technologies for Reducing Pesticide Use*, Wiley, NY. **1997**, 444 pp
4. Benbrook, C.M. *Pest Mangement at the Crossroads*. Consumers Union, Yonkers, NY. **1996**.
5. *Reducing Reliance,* ed. M. Watts; Macfarlane, R. Pesticide Action Network, Asia and the Pacific, Malaysia, **1997**, 93 pp.
6. Flora, C. *Plant Disease Reporter* Feb, **1990**, 105.
7. EPA, *Ecological Risk Assessment Issue Papers*, Risk Assess. Forum, Wash DC. EPA/630/R-94-009, Nov, **1994**.
8. Hall F. *ASAE,* ASAE Mtg, Phoenix AR., Paper No. 96104, **1996**.
9. Levitan, L. *Background paper of OECD*, **1997.** Copenhagen , DK
10. Marz, Uln, *Farming Systems and Resource Economics*. #7, **1990,** 245 pp.
11. Reichelderfer, K.; *pers com. **1997**.*
12. Hall, R. In *Novel Approaches to Integrated Pest Management*, ed, Reveni, CRC Press., Boca Raton, FL. ***1996***.
13. Stern, P.; Fienberg, H. eds, *Understanding Risk:Informing Decisions in a Democratic Society,* Nat Acad. Press. Wash DC, **1996**, 249 pp.

14. Jaenicke, E.C *Policy Studies Report No 8.*, Henry Wallace Inst. For Alternative Agriculture, Greenbelt MD, **1997**.
15. Parkin, C.; Blackmore, B. nespal.cpes.peachnet.edu/pf/ Web Site: http:// nespal/cpes.peachnet.edu/pf/., **1995**.
16. Sudduth, K.; Kitchen, N.; Hughes, D.; Drummond, S. In P. Roberts, et al., ed. *Site-Specific management for agricultural systems.* ASA Misc. Pub. ASA, CSSA and SSSA, Madison WI, **1995**, 671.
17. *The State of Site Specific Management for Agriculture*. Pierce, F.; Sadler, E.; Eds, Symposium Soil Science Soc. Am. and Am. Agron. Soc. Madison WI, **1997**, 430 pp.
18. *Precision Agriculture in the 21st Century.* Nat.Res.Council, Nat. Acad. Press. Wash., DC .**1997**, 490 pp.
19. Morgan, M.; Ess, D.; *The Precision - Farming Guide for Agriculturists.* John Deere Publishing. Moline, IL. **1997**, 117 pp.
20. Anderson, N.; Humbug, D. In, State of Site Specific Management for Agriculture, Pierce, F.; Sadler, E.; eds. Symposium Soil Science Soc. Am. and Am. Agron. Soc., Madison, WI, **1997**, 245.
21. Monson, R. ASAE, St Josephs, MI, ASAE Paper 96-1023, **1996**.
22. Hoffmann-Wellenhof. B.; Lichennegger, C.; Collins, J. *GPS:Theory and Practice*. Springer-Verlag, NY, **1994**.
23. Hanan, S. *The design and analysis of spatial data structures.* Addison-Wesley, NY, **1990**.
24. Tofte, D.; Hanson, L. ASAE, St Joseph, MI, ASAE, **1991**, Pub 11-91.
25. Anderson. N. ASAE, St. Josephs, MI. ASAE , **1993**, Paper: 93-3075.
26. Kosko. B. 1992. *Neural networks and fuzzy systems.* Prentice-Hall, Englewood Cliffs, NJ., **1992**.
27. Hall, F. *Phytoparasitica* 25 (Suppl), **1997**, 39S.
28. Ganzlemeir, H. 1998- IUPAC Invited Plenary – Pesticide Application, In press, **1999**.
29. Tompkins, F.; Howard, K.; Mote, C.; Freeland, R. 1990. Trans. ASAE, **1990**, 33, 439.
30. Stover, B. *Farm Chemicals.* March **1998,** 28.
31. Miller, M.; Smith, D.. 1992. Trans ASAE, **1992**, 35, 787.
32. Rew, L.J.; Miller, P..c.H.; Paice, M.E.R., l997. *Aspects of Applied Biology*, Optimising Pesticide Application, **1997**, 48, 49.
33. Giles, K. B.; *Atomization and Sprays*, **1997**, 7, 161.
34. Cardina, J.; Webster, T.; Herms, C.; Regnier, E., *Jnl Crop Production*, **1999**, *2, No. 1 (#3) 239.*
35. Hewitt, J.T.; Smith, K.R., In *Consortium News*, Consortium for Sustainable Agriculture Research and Education. Madison WI. April, **l996,** 9, 5.
36. Alessi, S. *Consortium News*, Sus. Ag. Research and Ed., **1996,** 9, 5.
37. Ventura, S. *Consortium News*, Sus Ag. Research and Ed., **1996,** 9, 5.
38. Wossink, G.; de Buck. A.; van Niejenhuis, J.; and Haverkamp. H., *Ag Systems*, **1997**, 55, 409.
39. Lowenberg-DeBoer, J. *Purdue Agricultural Economics Report*, Nov., **1997**.

40. Sulecki, J. 1997. *Farm Chemicals*, **1997**, Dec., 62.
41. Hendley, P.; Travis, K.; Homes, C.; Henrikson, E.; Kay, S. 1998 *IUPAC paper* –Probalistic aquatic risk assessment of pyrethroids. London, UK, 1998.
42. Hendley, P,; Laskowski, D.; Nelson, H.. 1998 *IUPAC Paper* –Regulatory exposure modeling in the USA – current status and outlook. London, UK.,**1998**.
43. Gustafson, E. *Ecosystems*, **1998**, 1, 143.
44. Lang, L , *Managing Natural Resources with GIS,* Environ. Systems Res. Inst., Cal, **1998**, 117 pp.

Chapter 9

Human Exposure Assessment in Risk Perception and Risk Management

R. I. Krieger

Personal Chemical Exposure Program, Department of Entomology, University of California, Riverside, CA 92521

Although there is concern from many sectors of society about health and risks associated with pesticides and other chemical technologies, there is little appreciation for the magnitude of unintentional, unavoidable, and accidental exposure. Even in cases when basic recognition of dose-response relationships seems evident, linear "zero" exposure extrapolations often result in predictions of harm from trivial chemical contact. Risk assessment may be a means to promote more objective evaluation of chemical exposures. Biological monitoring can clarify the extent of chemical absorption associated with particular activities which include chemical contact. Measurement of exposure can establish a basis for developing perspective. Barriers which limit pesticide uptake by mixer/loader/applicators include label uses, engineering controls, personal protective clothing, worker protection standard clothing, personal hygiene, and dermal absorption. Post-application exposures of harvesters and persons contacting residual pesticides sprays on treated indoor surfaces also have exposures related to source strength and activity. Recognition of the magnitude of these recurring, no-effect level exposures of very substantial numbers of people may contribute to the development of more balanced views of the significance of pesticide exposures of the general public that are associated with the diet, water, air and non-occupational activities.

Risk reduction in the use of pesticides is an important concept that is defined differently by regulators, manufacturers, product representatives, pest control advisers and operators, pesticide handlers, harvesters, retailers, consumers and their advocates, and politicians. Common risk reduction strategies include engineering controls, substitution of less toxic products, personal protective equipment, behavioral modification, regulatory and institutional responses and protective clothing. By developing of more accurate personal chemical exposure data, the risk of exposure to chemicals used as pesticides can be reduced. Without modifying pesticide use practices or using "safer" pesticides, more accurate human exposure data can result in risk assessments that are less reliant upon default assumptions that inflate exposure assessments and may misguide development of mitigation measures.

Pesticides are virtually never used in a pure form. Whether isolated from an aboriginal poultice, the tars of a synthetic organic chemist, or a biochemist's broth, a pesticide active ingredient is one chemical among many chemicals. Discovery of beneficial killing activity of a substance results in the formulation of an active ingredient in a more simple and better defined matrix used to deliver the active to its target. No other economic class of chemicals is so extensively used to minimize the impact of our competitors for food and fiber, shelter, and vectors of disease.

Current concerns about *risk*, the probability of an adverse effect resulting from chemical exposure, is driven by the reality that pesticide use is inevitably and inextricably linked with human exposure. In the overwhelming majority of cases, these exposures are similar in magnitude to those attributable to natural products or artifacts of other chemical technologies and sufficiently low to be benign to health. Except in rare accidental episodes, most unintentional and unavoidable human exposures to active ingredients and their derivatives result from pesticide persistence and distribution of very small amounts in human environments. Because of the inherent toxicity of pesticides to pests and a poorly informed and confused public, immense resources are annually committed to documenting the fate and transport of some pesticides in a variety of matrices including air, water, soil, plants and producc, foods and fccds, wildlife and, to a much lesser extent, humans.

Success or failure of exposure reduction measures are judged by using a toxicological standard- the no observed adverse effect level (NOEL)--resulting from hazard identification and dose-response studies in animals. Minimizing risk resulting from absorption of chemicals used as pesticides entails reducing or mitigating *exposure* relative to the NOEL. Exposure is the measure of the environment leading to a dose. It is measured as the concentration of a chemical in the matrix in contact with an organism (human), integrated over the duration of the contact (*1*). Absorbed dose (internal dose) is the amount entering systemic circulation after crossing a specific barrier such as skin, lung, or digestive tract. Absorbed dose is the unit of measure used in calculating margin-of-exposure (margin-of-safety), the ratio of the NOEL to absorbed daily dosage (ADD, mg/kg/day). Although both the numerator and denominator of the margin-of-exposure (margin-of-safety) expression commonly include default assumptions resulting from incomplete knowledge, the accuracy of pesticide exposure assessments can be improved so that apparent risk can be reduced.

Organophosphates are particularly useful tools to study the relationship between use and human pesticide exposure because of their many patterns of use, similar product chemistry, and well-characterized disposition in humans and the environment. The rapid clearance of metabolites in urine, that are stable biomarkers of oral, dermal, or inhalation exposure, is particularly important to exposure monitoring.

This paper discusses pesticide exposure data collected in the Personal Chemical Exposure Program, Department of Entomology, University of California, Riverside, on unintentional or unavoidable exposures from the normal use of registered products of pesticide handlers (mixer/loader/applicators), harvesters entering treated fields, and persons living in pesticide-treated residences. The exposures resulted from organophosphate use as a dormant spray, organophosphate insecticide use in protection of row crops, and finally organophosphate use in control of flies and fleas in California homes.

Methods.

The approach for determination of exposure and pesticide clearance is similar in each case: establish the insecticide to be used, where it will be used, and identify the population of people that will be exposed. Careful consideration of the nature of human exposure will usually result in selection of an appropriate means to measure a biomarker of ADD. In these cases with organophosphates, measurement of blood and plasma cholinesterases or urine biomonitoring were options, but urinary metabolites are unquestionably more accurate for sensitive and specific analysis at normal levels of exposure.

Biomarkers of Exposure and Absorbed Dose. Several organophosphate insecticides were used as indicators of human exposure. They include chlorpyrifos, diazinon, methidathion, naled, and malathion. Their metabolism at non-toxic, low doses in humans is similar. Urine clearance of the common alkyl phosphates (AP) usually includes more dialkylthiophosphate than dialkylphosphate. Des-alkyl products have not been measured at normal levels of exposure. It seems likely that hydrolysis is quantitatively more important than desulfuration in biomonitoring. Unfortunately, the dialkylphosphate esters are detectable at about 25 ppb in urine making them unsuitable for most exposure monitoring. The vast majority of human (e.g. 60 kg) exposures will result in clearance of dosages less than 1 ug equivalent organophosphate (FW OP$\approx$ 300) per kg body weight (25 ppb AP x $2_{FW\ OP}$/1.5 L urine/60 kg).

Malathion yields the expected mono- and diacids as well as the corresponding dimethyl phosphates that have considerable utility for human monitoring. Detection limits for the acids are 2-4 ppb, about an order of magnitude below the limits for the dimethyl products produced in lower amounts and with higher detection limits. The metabolites are stable in urine and cleared rapidly following dermal, inhalation or oral exposures.

Leaving group analysis facilitates determination of the estimation of chlopyrifos, methyl- and ethyl parathion, and diazinon. The leaving groups, trichloro-2-pyridinol, p-nitrophenol and 2-isopropyl-6-methyl-4-pyrimidinol, are excreted as conjugates that must be hydrolyzed before analysis. Recent illegal methyl parathion use and follow-up remediation efforts were guided, in part, by urine biomonitoring (*2*). However, because the leaving groups themselves are also produced in the environment (*3*), their detection in urine may not always be associated with exposure to the parent insecticide.

Biomonitoring permits investigators to quantitatively relate human exposure to experimental dose-response studies. The trichloro-2-pyridinol can be measured in human urine at about 4 ppb (limit of quantitation, LOQ). If we assume a urine production of 1.5 L/day for a 70 kg male, the estimated absorbed dose at the LOQ for the method would be about 6 ug or the ADD would be 0.1 ug/kg. That will provide an ample range between the default of the dosage determined by 1/2 the LOQ, NOEL, the lowest observed adverse effect level (LOEL), and the lethal oral dose to 50% of an experimental rat population (LD50).

Pesticide Handlers. Workers who mix, load, and apply have the opportunity for high exposure and associated risk. Their work has been safened during the past 50 years by many innovations related to pesticide formulation (e.g., powders vs. granules vs. liquids vs. water-soluble sachets), closed transfer technologies, improved hose fittings and couplings, application methods, personal protective clothing, plus additional methods and techniques implemented to reduce risk. It remains true that the exposure reduction potential of most of these procedures has not been assessed using estimates of absorbed dosage (*4*), but their positive impact on the workplace is clearly evident in modern agriculture, urban pest management, and vector control.

Initial efforts to assess human exposure were guided by the reality of over-exposure and organophosphate toxicity in handlers and harvesters. Griffiths et al. (*5*) monitored the inhalation exposure of parathion applicators using respirator filter traps. Shortly thereafter, Bachelor and Walker (*6*) reported potential dermal exposure after analysis of pads affixed to clothing during routine work activities. The critical studies of Durham and Wolfe (*7*) revealed means to measure potential exposure and demonstrated the importance of dermal exposure.

Many studies of worker exposure followed and results are represented in databases such as the Pesticide Handlers Exposure Database (*8*). The database is a useful tool but requires additional information to achieve reasonable estimates of ADD. Potential exposures estimated by the use of cotton patches inflate potential dermal absorption (PDE) because of their placement, size, and propensity to retain pesticide spray particles. PDE must be factored by clothing penetration and dermal absorption to yield a reasonable estimate of exposure and risk. The clothing penetration problem may be overcome by use of a whole body garment, e.g. union suit or "long johns," as a dosimeter beneath the work clothes. For the purpose of estimating ADD both clothing penetration (default 10%) and dermal absorption must be weighed-in to permit calculation of ADD from the PHED.

Clothing Penetration. Work clothes are an important protective barrier against direct skin contact. Early investigators used inner and outer cotton gauze patches to assess the penetration of pesticides in handlers. Recently, investigators at the California Department of Pesticide Regulation have analyzed registration data by regression analysis. Their work using a large sample of pesticide registration data confirms the protectiveness of clothing. Their analysis also shows that low rates of application result in greater clothing retention than higher rates.

The accuracy of PHED-projected worker exposures and the protectiveness of outer garments was recently evaluated in workers who applied dormant oil-organophosphate sprays using air blast equipment in the Central Valley of California (*9*). Handlers wore either Tyvek®-Saranex® and Kleengard LP®, and urine biomonitoring was used to measure alkyl phosphate clearance of the handlers who applied organophosphate-dormant oils. The barrier properties of the garments with respect to day-to-day exposure were identical during a 2-week study period. The ADDs ranged from 6 ug/kg to 8 ug/kg based upon the equivalents of chlorpyrifos, diazinon, methidathion, and naled applied using air-blast applicators. These ADDs are less than those predicted by the PHED database even when the model is adjusted for clothing penetration (10%) and default dermal absorption (10%).

Body Weight. Before consideration of harvester exposures, an additional factor of importance to both handlers and harvesters is the denominator of the expression of ADD (ug/kg bw). Body weight is too often forgotten or overlooked in estimating ADD. Persons of different body weights participate in exposure monitoring work. These exposures are ultimately expressed as dosage so the estimates can be related to dosages from animal studies as part of the risk characterization process. Although there is likely a relationship between body weight, surface area, and absorbed dose, there is no evidence for determination of absorbed dose that any factor is more important than personal behavior or work practices. The unit of exposure is the individual, and when feasible absorbed, dose should be expressed on a per person basis or normalized by measured body weight. When workers participate in exposure studies, it contributes to accuracy to record actual body weights, rather than to assume when the final report is being prepared that the 70 kg default man, 50 kg female or 60 kg person should be applied. A recent series of weighings of pest control operators (PCO) in California provided a memorable example of the importance of measured body weight. The workers who volunteered for the study were attending periodic PCO training meetings sponsored by Target. Use of the 70 kg default body weight for this group would result in a 29% overestimate of the absorbed daily dosage (mg/kg). It is difficult to conceive of a single factor in the exposure algorithm that can so significantly effect ADD.

Harvesters of Treated Crops. In the 1950s excessive organophosphate exposures, particularly ethyl parathion, occasionally resulted in acetylcholinesterase inhibition and poisoning of workers harvesting treated crops. In other cases episodes of contact dermatitis were attributed to contact with pesticide treated crops. These acute illnesses resulted in a protective system of reentry intervals on a specific crop/chemical basis. More recently, reproductive and developmental toxicity and chronic exposures have become concerns. Since work clothes and the time interval between pesticide application and significant worker contact with treated foliage are the most effective means to protect harvesters from excessive exposure, both are important elements in the development of contemporary exposure-based entry.

Exposure-based entry intervals represent an important application of risk reduction strategies (*10*). The empirical relationship between dislodgeable foliar residues (DFR, ug/cm^2) and dose initially described by Nigg et al. (*11*) and Zweig et al. (*12*) (see Figures 1 and 2). Potential dermal exposure (*13*) or hourly dermal exposure (DE_h, ug/hour) results from transfer of pesticide from treated foliage to outer garments. Transfer to exposed skin (dermal exposure) results from clothing penetration or direct contact with skin. Transfer factor (TF, cm^2/hour) was derived from exposure data collected from harvester exposure monitoring. Under most circumstances inhalation and ingestion are considered negligible for the establishment of protective entry intervals. This PDE (or hourly dermal exposure) can be expressed in the following equation:

$$\text{PDE or } DE_h = \text{DFR x TF}$$

Appreciation of the importance of the extent of contact transfer resulting from particular work tasks has improved the usefulness of this means for estimating worker exposure. To estimate absorbed daily dosage (ADD, ug/kg hours per day (H/day), for contemporary risk assessment, measures of the protectiveness of clothing (10% penetration), dermal absorption (ABS, %/24 hours), and body weight (kg) must be introduced and can be calculated as:

$$\text{ADD (ug/kg /day)} = (DE_h \text{ x ABS x H/day)/kg bw} = \text{(TF x DFR x ABS x H/day)/kg}$$

From this equation, the estimated ADD_{SL} at the Safe Level (SL) or the DFR_{SL} can be estimated as:

$$ADD_{SL} = \text{[(TF x ABS x H/day)]/kg bw x } DFR_{SL}$$

or

$$DFR_{SL} = (ADD_{SL} \text{ x kg)/(TF x ABS x H/day)}$$

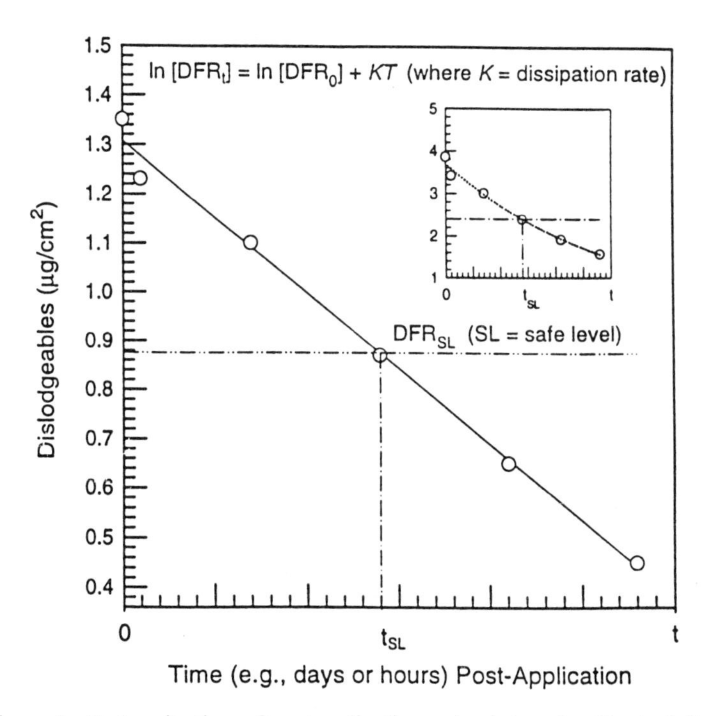

Figure 1. Determination of postapplication entry intervals. Typical first order decay of natural log dislodgeable foliar residue (DFR). Inset uses linear ordinate.

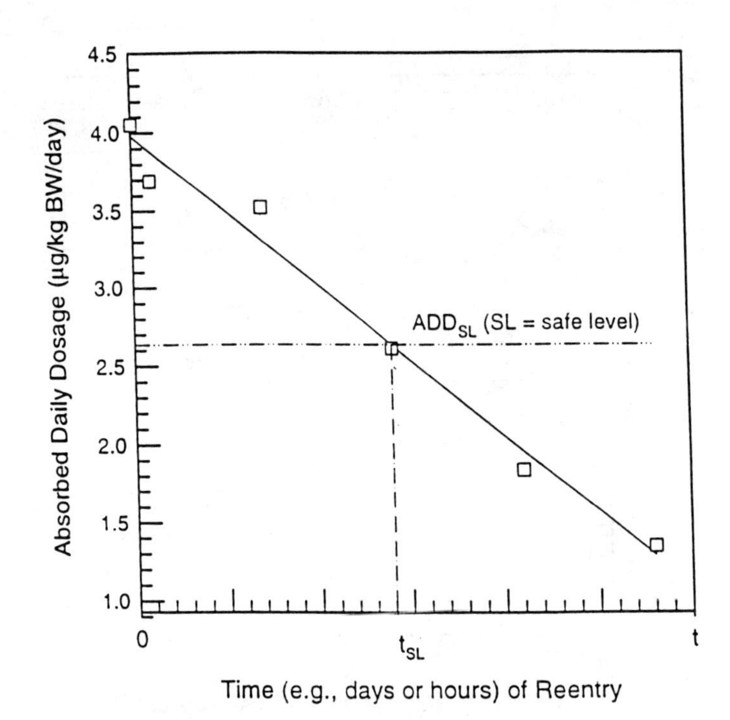

Figure 2. Decline of estimated absorbed dosage. Natural log of absorbed daily dosage at a safe level identifies time of safe reentry. Dong and Ross, California Environmental Protection Agency, personal communication.

In, the logarithmic decay of DFR (Figures 1 and 2) is linearized by the equation,

$$\ln [DFR_{SL}] = \ln[DFR_{t=0}] + Kt$$

and

$$t = \{\ln[(ADD_{SL} \times kg)/(TF \times ABS \times H/day)] - \ln[DFR_{t=0}]\} \times K^{-1}$$

In 1995 and 1997 malathion exposures of strawberry harvesters at Watsonville, California, were monitored. The mono- and diacids of malathion were monitored in 24-hour (1995) and morning (1997) urine specimens. The DFRs were 0.15 ug/cm^2 and 0.064 ug/cm^2, respectively (Table 1). The measured TFs were calculated as 2,053 and 1,834 cm^2/h.

Hands as a Route of Exposure. Hands account for only 5.2% (range 4.6-7.0%) of the body surface area of adult males (14), yet they are singularly important in the dermal absorption of pesticides in the workplace. Recognition of their importance as sources of pesticide exposure is not new (*7*), but the estimation of the quantitative contribution of hands to absorbed dose is very poorly studied.

Most estimates of hand exposure are based upon the relatively large amounts removed from the hands by rinsing, washing, or absorbent gloves. Hand rinses are collected in various solutions ranging from aqueous surfactants to neat isopropanol or ethanol, depending upon the physicochemical properties of the analyte. Similarly, cotton gloves may be worn as the work (contact) surface or beneath protective gloves during normal activities to sample potential dermal exposure. Pesticide residues retained on the skin but unavailable for dermal absorption results in overestimates of dermal absorption. Examples of unavailable (or very poorly available residues) are chemicals bound to soil or vegetable matter and layers of residue which frequently accumulate on the back of the hand, arms, and V of the neck.

In recent experiments we have more directly assessed the contribution of hands to absorbed dose in strawberry harvesters. Krieger (*13*) reported that about a 50% reduction in absorbed malathion occurred in harvesters who used rubber latex gloves compared to absorption by bare handed workers in the same fields. An isopropanol (50% v/v) rinse removed about 3 to 10 times more malathion from workers' gloves than was absorbed by the ungloved workers. This finding provides direct evidence of the contribution of hands to absorbed dose, and the high pesticide levels recovered from glove rinses make gloves an unreliable dosimeter under most conditions. During the 1997 growing season, harvesters applied a newly-developed skin protectant lotion to their hands before work began. Urinary clearance of malathion metabolites (24 hour) was reduced in the persons using the protective lotion (Krieger et al., unpublished). These interesting preliminary findings warrant more complete study with respect to workplace hygiene and risk reduction.

Table 1: Estimated Malathion Transfer Factor for Strawberry Harvesters

Year	DFR ug/cm^2	Calculated ADD[1] ug/day	Measured ADD[2] ug/day	TF[3] cm^2/h
1995	0.15	295	202	2,053
1997	0.064	126	77	1,834

[1] DFR x 8 x 3,000 (Cal-EPA TF) x Dermal absorption
[2] Estimated by urine biomonitoring, Krieger et al.(unpublished)
[3] TF = Measured ADD/ (Dermal absorption x 8 hours x DFR)
Dermal absorption of malathion = 8.2%/24 hours (Thongsinthusak, Personal Communication)

The well-established tendency of hands to accumulate environmental residues has been amply documented by previous rinse and wash-off procedures as well as studies in which glove adsorption is used for sampling. In two field studies with strawberry harvesters, we have begun to clarify the importance of hands as a site of dermal absorption. Most frequently, harvesters wear gloves for protection from the elements as well as workplace safety rather than as a means to reduce pesticide absorption. On the basis of the utility of handwash procedures using dilute detergent to estimate dermal loading, routine handwashing with soap and water would probably satisfy the needs for personal hygiene and cosmetics as well as reduce dermal pesticide residues.

Residences Treated With Foggers or Area Sprays. Berteau and Mengle (*15*) of the California Department of Health Services and Maddy of the Department of Food and Agriculture conducted preliminary review of pesticides used indoors. They noted several cases (6) from the California Pesticide Illness Surveillance System in which illness were reported after structural pest control. On the basis of review of pesticide use practices and the nature and time to onset of symptoms, the cases likely resulted from sensory (especially odor) responses and confusion about the nature of the responses rather than from systemic toxicity. Subsequently, the hypothetical exposure estimates were developed by the California regulators for infants, children, and adults after label use of propoxur, DDVP, and chlorpyrifos were sometimes greater than toxic levels. Berteau et al. (*16*) reiterated their concerns, particularly for children.

Considerable attention and effort by scientists in academia, regulatory agencies, and the industrial sector followed discussion and publication of the default indoor scenarios of the late 1980s (*16*). During the same interval the development of formal pesticide risk assessments became a much more common practice (*17*), and a premium was placed upon human exposure data. The more recent Food Quality Protection Act of 1996 brings increased attention to monitoring human indoor and outdoor post-application pesticide exposures. In marked contrast to the previous circumstances to develop safe pesticide use practices for handlers and harvesters, this current effort has been pursued with virtually no knowledge of the sample frame of human exposure. As a result the numerous methods for monitoring indoor air and surfaces that have been developed bear uncertain relationship to the dynamics of indoor human pesticide exposure.

An experimental evaluation of the exposure potential of indoor foggers was initiated shortly before Berteau et al. (*16*) published their alarming (mg/kg) default exposure estimates. A carefully controlled, 20-minute series of high-contact activities (Jazzercise™) were selected for use by persons wearing cotton, whole body dosimeters (socks, gloves, and union suits). As a result of high contact during two 20-minute periods and efficient transfer of pesticide to the cotton dosimeter, a person's indoor daily (24-hour) dermal pesticide exposure has been approximated. The experimental protocol has remained virtually unchanged during the past 10 years (*18*), and more data have been obtained for further validation (Krieger et al., In Press).

Perhaps more important results have been obtained from a series of monitoring and experimental studies in the Personal Chemical Exposure Program, Department of Entomology, University of California, Riverside. Situational (opportunistic) monitoring of residents who made routine use of pesticide foggers and area sprays as well as controlled experimental studies have been performed. The normal range of human daily pesticide exposures is ug/kg rather than mg/kg (*16*). Preliminary review of these data reveals that children may absorb 3-10 times more insecticide than their parents who spend similar amounts of time indoors.

Of equal interest is the duration of pesticide exposure after indoor use (Figure 3). Three important experimental indoor studies have been performed using foggers. In this instance, the mean daily chlorpyrifos exposures (± SD) were estimated by biomonitoring for urinary clearance of TCP. Seven family members (aged 18 to 88) lived 16-24 hours per day in the 10 rooms of their 1,500 square foot, two-story home in southern California. Six flea foggers discharged 1.8 g chlorpyrifos, resulting in about 5 ug/cm^2 (about 4 feet from the fogger) on carpeted surfaces within the home. After a 3-day study in 1996, urine specimens were collected for a more extended period in 1997 (Figure 3). The results were not predicted by our own environmental (air and surface) monitoring (*18, 19*) or air levels reported in the literature (Figure 4). Environmental levels decline much more rapidly (hours) than human exposure potential (measured in days). Urine clearance of TCP was maximal during the first week, reflecting the 27-hour half-life of chlorpyrifos in humans (*20*). After 1 month, daily TCP clearance had not returned to control levels. The extended duration of low-level indoor human exposure indicates that measurements of air and surface pesticide levels as they are presently made are not predictive of indoor post-application exposure.

Summary

Analytical chemistry has established methods and techniques that permit routine measurement of pesticides and their metabolites in human and environmental samples. The resulting measurements form the foundation for exposure assessments that are rudiments of the risk characterization process. These measurements are not easily transformed into absorbed dose. Undue reliance upon default assumptions associated with patterns of pesticide use, chemical stability, and bioavailability may magnify dose and result in irresponsible estimates of exposure and risk. Although failure to recognize health risks may unnecessarily expose people to hazardous conditions, the present climate seems more likely to spawn careless extrapolations and unfounded, hand-wringing concern that perhaps is a greater threat to health and well-being than environmental levels of chemicals. Future more accurate

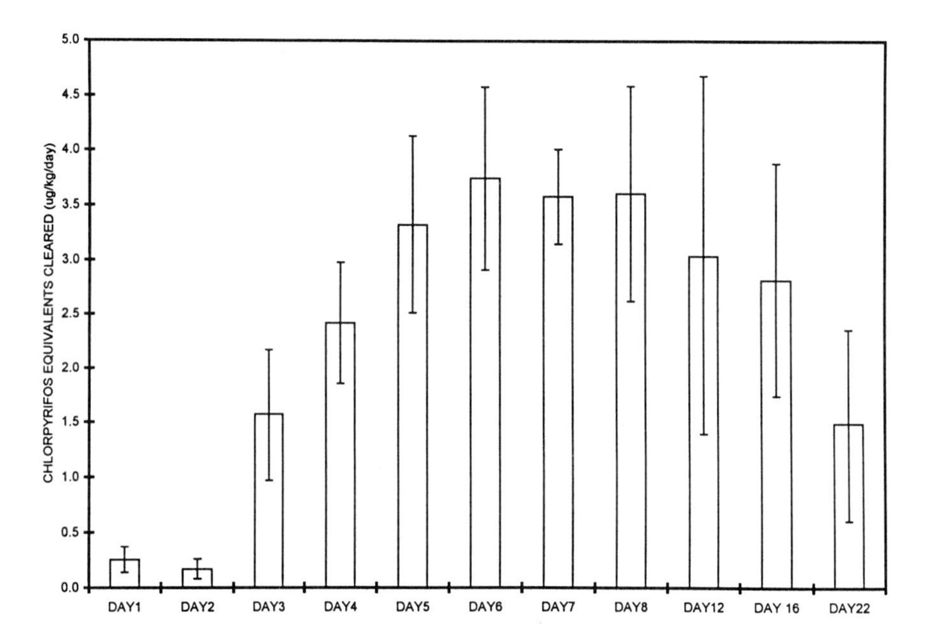

Figure 3. Mean urine clearance of trichloropyridinol (±SD) by seven adults (aged 18 to 88) in a 2-story southern California home (ca. 1,600 ft^2). Day 1 was 24 hours prior to fogging of with six 1.8 g foggers. Foil coupons placed about 4 feet from cannisters retained about 5 ug chlorpyrifos/cm^2. The family reentered after 2-hour application and 30-minute ventilation period. Morning voids were collected each day and analyzed for TCP. Absorbed dosage was not corrected for background, but volume was adjusted for creatinine, assuming 1 and 1.7 g creatinine/day for women and men, respectively.

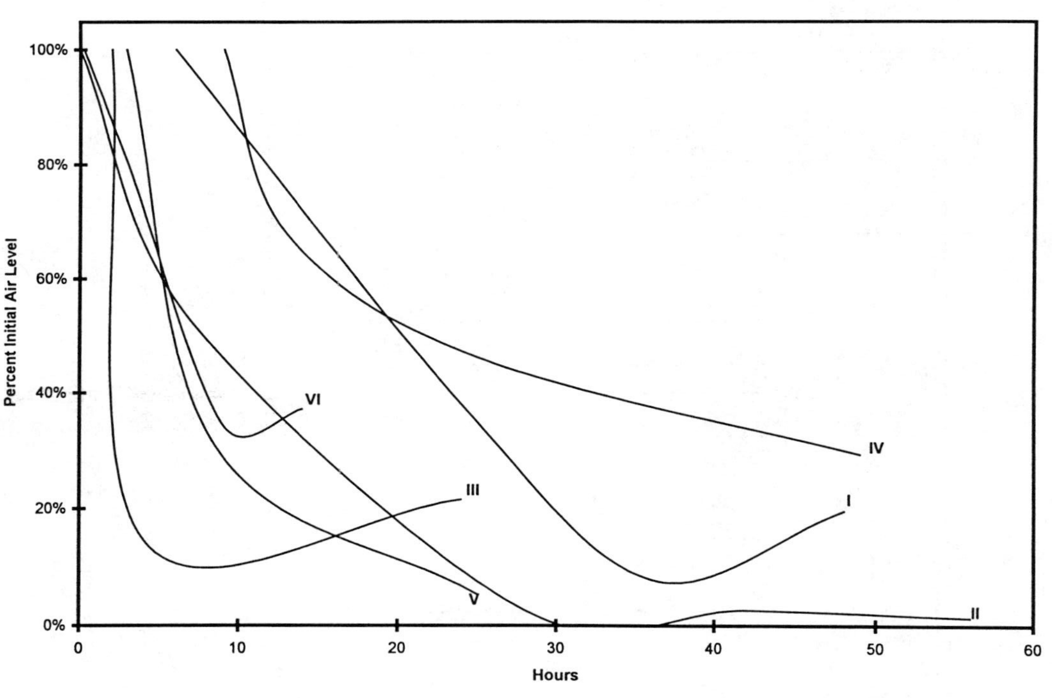

Figure 4. Air levels after indoor uses of chlorpyrifos. Air levels have been the most frequently measured environmental indicator of potential exposure to pesticide vapors and particles. After short intervals, air levels drop sharply. In the cases shown here, markedly different levels of chlorpyrifos in air were measured during the period shown (terminated at 60 hours because few measurements have been reported beyond that period). I: (*21*); II: (*22*); III:(*23*); IV:(*24*); V:(*25*); VI: (*26*).

exposure assessments will be founded upon a thorough understanding of patterns of use, knowledge of exposure and physiologic disposition, and more meaningful sampling of the physical and chemical environment.
Human biomonitoring provides opportunities to assess collective (or aggregate) pesticide exposure. Biomonitoring is the best available reality check on estimates derived from the analysis and summation of apparent environmental sources of exposure. Biomonitoring can reduce risks by contributing additional information to the risk assessment process as well as providing data for development of biochemical exposure indices as part of health surveillance programs.

Acknowledgement Research support of the Division of Agriculture and Natural Resources, University of California, and the California Strawberry Commission are gratefully acknowledged. Michael Dong and John Ross, Worker Health and Safety Branch, Division of Pesticide Regulation, Cal-EPA contributed discussion on reentry. Gratitude is expressed for the continued cooperation of area families in residential monitoring work. Administrative assistance ably provided by Sylvia Garibay.

Literature Cited

1. U. S. Environmental Protection Agency. 1988. Federal Register 53: 48830-48853.
2. Esteban, E.; C. Rubin, R.; Hill, D.; Olson, K.; Pearce. J. Exp Anal Environ Epidemiol (1996) 6: 375-367.
3. Racke, K.D. Rev. Environ. Contam. Toxicol. (1993), 131: 1-154.
4. Rutz, R.; Krieger, R.I. In: G.W. Ware (Ed.), Reviews of Environmental Contamination and Toxicology, Springer-Verlag, New York, Vol. 129, 1992, pp. 121-139.
5. Griffiths, J. T.; Stearns, Jr. C. R.; Thompson, W. L. J. Econ. Entomol. 1951, 44: 160-163.
6. Batchelor, G. S.; Walker, K. C. AMA Arch. Ind. Hyg. 1954, 10: 522-529.
7. Durham, W.F.; Wolfe, H.R. *Bull. WHO* 1962, 26: 75-91.
8. PHED (1992) Pesticide Handlers Exposure Database, USEPA, Health and Welfare Canada, National Agricultural Chemicals Association.
9. Krieger, R.I.; Dinoff, T.M.; Korpalski, S.; Peterson, J. Bull. Environ. Contam. Toxicol. (1998) 61: 455-461.
10. Maddy, K.T.; Krieger, R.I.;. O'Connell, L; Bisbiglia, M.; Margetich, S. In: Wang, R.G.M.; Franklin, C.; Honeycutt R.C.; Reinert, J.C. (Eds.), American Chemical Society Symposium Series 382. American Chemical Society, Washington, D. C., 1989, pp. 338-353.
11. Nigg, H. N.; Stamper, J.H.; Queen, R. M Am. Ind. Hyg. J. (1984), 45: 182-186.

12. Zweig, G.; Leffingwell, J. T.;. Popendorf, W. L; Environ, J. Sci. Health (1985), B20: 27-59.
13. Krieger, R.I. In: Reed, D. R. (ed.) Proc. Intl. Cong. Toxicol.-VII, Elsevier, Amsterdam, 1995, pp. 65-72.
14. U. S. Environmental Protection Agency, Exposure Factors Handbook (1989), EPA/600/8-89/043.
15. Berteau, P. E.; Mengle, D. M. Community Toxicology Unit, California Department of Health Services, Berkeley, May 17, 1985.
16. Berteau, P. E.; Knaak, J. B.; Mengle, D. C.; Schreider, J. B. In: Wang, R.G.M.; Franklin, C.; Honeycutt, R.C.; Reinert, J.C. (Eds.), American Chemical Society Symposium Series 382. American Chemical Society, Washington, D. C., 1989, pp. 315-326.
17. National Research Council, National Academy of Sciences. Risk Assessment in the Federal Government: Managing the Process. National Academy Press, Washington, D.C., 1983, pp. 1-191.
18. Ross, J.; Thongsinthusak, T.; Fong, H.R.; Margetich, S.; Krieger, R. Chemosphere 1990, *20*, 349-360.
19. Ross, J.; Fong, H. R.; Thongsinthusak, T.; Krieger, R.I. In: Guzelian, P. S.; Henry, C. S.; Olin, S. S. (Eds.), Similarities & Differences Between Children & Adults, ILSI Press, Washington, D.C., 1992, pp. 226-241.
20. Nolan, R.J.; Rick, D.L.; Freshour, N.L.; Saunders, J.H. Toxicol. and Appl. Pharmacol. 1984, 73: 8-15.
21. Lu, C.; Fenske, R.A. Air and Surface Chlorpyrifos Residues Following Residential Broadcast and Aerosol Pesticide Applications. (In press).
22. Leidy, R.B.; Wright, C.G.; Dupree, H.E. Jr. Environ. Monitoring and Assess 1996, 42: 253-263.
23. Bohl, R. W. Unpublished report, The Dow Chemical Company, 1983.
24. Koehler, P.G.; Moye, A.H. J.Econ. Entomol. 1995, 88(4): 918-923.
25. Vaccaro, J.R. Risks Associated with Exposure to Chlorpyrifos and Chlorpyrifos Formulation Components. Amer. Chem. Soc., 1993, pp 297-306.
26. Ross, J.; Fong, H.R.; Thongsinthusak, T.; Margetich, S.; Krieger, R. Chemosphere 1991, *22*, 975-984.

Chapter 10

Impact of Patent Policy on Bioremediation with Living Organisms

H. N. Nigg [1], D. R. Saliwanchik [2], and R. Saliwanchik [2]

[1] University of Florida, 700 Experiment Station Road, Lake Alfred, FL 33850
[2] Saliwanchik, Lloyd, and Saliwanchik, 2421 N.W. 41st Street, Gainesville, FL 32606-6669

Bioremediation with living organisms and genetic engineering began with the 1977 U.S. Court of Customs and Patent Appeals decision in favor of Berge, Malik and Coats' lincomycin process patent. This decision lead to a favorable decision for the petroleum digesting microbe of Chakrabarty (General Electric). These decisions were upheld in 1980 by the U.S. Supreme Court. As a result, living organism patents have increased from 1 in 1974 to 20 in 1997/98 and a new and vibrant industry was created.

Bioremediation has become an enormous scientific and commercial field. For example, as of this writing there were about 1350 pages on the world wide web on bioremediation. For the scientific and market details of bioremediation we suggest references *1-26*. Patents have played an important role in facilitating the development and commercialization of bioremediation technology and other areas of biotechnology. From its origins in the 1700's until now, the patent system has promoted scientific progress by providing public access to complete details of emerging technologies and by rewarding the creators of useful innovations. The recognition of the patentability of living organisms is a testament to the ability of the patent system to adapt to changing times and evolving technology. The court decisions establishing the patentability of living organisms have opened the door for legal protection of research which has led to products and processes which benefit society by providing a cleaner environment through bioremediation.

General Overview of Patent Process and the Requirements for Patentability

The Congress shall have power...To promote the progress of science and useful arts, by securing for limited times to authors and inventors the exclusive right to their respective writings and discoveries (*27*).

"Ingenuity should receive a liberal encouragement." V. Writings of Thomas Jefferson (*28*).

Whoever invents or discovers any new and useful process, machine, manufacture, or composition of matter, or any new and useful improvement thereof, may obtain a patent therefor, subject to the conditions and requirements of this title. 35 United States Code §101.

Congress intended [patentable] subject matter to include anything under the sun that is made by man: S. Rep. No. 1979, 82d Cong. 2d Sess. 5 (1952); H. R. Rep. No. 1923, 82d Cong. 2d Sess. 6 (1952) (*29*).

The above passages provide a framework for discussing the patenting of new classes of technology including, for example, plants, microbes, and other inventions adapted for improving our environment. The constitution of the United States contains a provision which provides Congress with the power to create a patent system to promote the progress of science (*27*). Thomas Jefferson's writings confirm that he, as one of the founding fathers of this country, believed in the importance of fostering human ingenuity (*28*). The federal statutes enacted to implement the constitutional provision calling for a patent system reflect the founding fathers' desire to promote ingenuity and the dissemination of information regarding new inventions (*30*). More recently, the Supreme Court of the United States has recognized and re-affirmed the broad role of patents in promoting the progress of science and mankind (*29*).

In recent years, many new technologies have been developed to prevent pollution; clean up polluted land, air, and water; make agriculture more productive; and reduce the need to introduce chemicals into the environment for agriculture or other purposes. Many of these new technologies exploit advancements and innovations in the biological and microbiological sciences. The success of these new technologies is both dependent upon, and reflected in, the patents which have issued for these technologies. In order for the patent system to provide protection for these inventions and to promote further research and investment, it has been necessary in some instances for the patent law to evolve with the emerging technologies. This co-evolution of patent law and technology is particularly evident over the last 25 years in the biotechnology field. For example, in the early 1980's the Supreme Court affirmed the decisions by the Court of Customs and Patent Appeals (CCPA) in the Bergy (*31*) and Chakrabarty (*32*) cases holding that living cells can be patented. In the Bergy case, the patent applicant (represented by R. Saliwanchik) successfully argued in favor of the patentability of "biologically pure cultures." The Chakrabarty decision acknowledged the patentability of cells which had been genetically engineered to confer upon those cells new and advantageous capabilities. These legal decisions are discussed in more detail below. First we provide an overview of the patenting process and the requirements of patentability.

Principles of the Patent System

When distilled to its most basic elements, the patent system is simply a means to encourage innovation and promote public dissemination of new ideas and discoveries. The founding fathers of our country included within the constitution of the United States a provision calling for patents and copyrights to "promote the progress of science and the useful arts."

The patent system is not designed to monitor the implementation of technology or evaluate environmental or social ramifications of the development of technologies. The risks inherent in the development or implementation of all new technologies should be carefully considered and weighed against the potential benefits of the technology. If the risk/benefit relationship is such that a new technology should be developed, then continued efforts should be made to minimize any potential risks. The analysis of the risk/benefit relationship, as well as the promulgation of regulations to ensure public health and safety, is carried out by trained professionals in government agencies such as the EPA, USDA, NIH, and FDA. This process of risk/benefit analysis and risk minimization should be carried out with the benefit of as much relevant information as possible. Thus, although the government has ultimate responsibility for many decisions relating to public health and safety, the scientific community, religious and academic leaders, and the general public all can, and should, provide informed input during this process.

In order to provide informed input, it is critical for these sectors of society to have as much access to up-to-date accurate technological information as possible. In this regard patents perform a critical function in providing public dissemination of state-of-the-art technological information.

Patents are granted only after the PTO has determined that an invention, and its patent application, meet the strict requirements for patentability which have been established by Congress. The Patent Office employees given the responsibility of reviewing patent applications and making patentability determinations are known as patent examiners. Each examiner has at least a bachelor's degree in some scientific field; many examiners have doctorates, are lawyers, and/or have significant work experience.

Each patent application received by the PTO is assigned to an examiner who is trained in the scientific field to which the invention pertains. The patent examiner reviews the application to ensure that all of the requirements relating to the form and the substance of the application have been satisfied. Of primary significance with regard to the content of the application is the requirement that the applicant provide a complete written description of how to make and use the invention (*33*). This description must be sufficiently detailed and complete so as to enable a person skilled in the art to make and use the invention without undue experimentation. Such a full, detailed description is known as an "enabling" disclosure (*34*). This complete detailed account of the invention is published when a patent is granted in the United States, and/or 18 months from the filing date if an international application is filed (*30, 35-37*). The publication of this description plays a central role in the patent system. Specifically, this publication enables other researchers and interested parties to have full knowledge of the technology so that they can improve on the technology and combine these teachings with their own knowledge and/or other such teachings, thereby efficiently expanding the store of human knowledge.

In addition to the written description of the invention, a patent application must include at least one "claim" (*33*). A claim is a concise statement, found at the end of a patent application, which succinctly states the subject matter which is to be covered by the patent. When a patent is granted, the patent holder can prevent others from making, using, or selling *only* the subject matter *covered by the claims*. The claims of a patent can never cover more than what has been enabled by the patent's description. If a patent is granted, the duration of the patent rights is 20 years from the filing of the application.

In addition to the requirements of the patent application, there are strict requirements on the characteristics of the inventions which can be patented. These requirements have been promulgated by Congress in order to ensure that patents are awarded only for inventions which are the result of human inventive ingenuity and which represent substantial advancements over anything which was previously known to man.

In the United States, there are three primary requirements which an invention must meet in order to be patentable. These are novelty, non-obviousness, and utility. Each of these is discussed briefly below.

Novelty. To be "new" under the patent laws, an invention must not have been known and available to the public prior to the time when the applicant for patent "invented" it (*38*). Accordingly, if an uninformed researcher were to independently "discover" penicillin today, a patent would not be awarded because isolated and purified penicillin is already known and in the public domain. Similarly, chemicals, cells, viruses or other entities which exist in nature prior to the date of invention *can not* be patented in their native form because they are not new.

Non-obviousness. The U.S. patent statutes express the non-obviousness requirement as follows: A patent may not be obtained though the invention [satisfies the novelty requirements] if the differences between the subject matter sought to be patented and the prior art are such that the subject matter as a whole would have been obvious at the time the invention was made to a person having ordinary skill in the art to which said subject matter pertains (*39*).

The purpose of the non-obviousness criterion is to prevent the granting of patents for inventions which are merely predictable and/or are small advances over known technology. Therefore, in order to satisfy the non-obviousness requirement, the patent applicant may need to demonstrate that the invention was unexpected, highly advantageous, or otherwise more than the next logical step in the course of research.

Utility. Another requirement for patentability is that the invention be useful (*39*). Accordingly, a chemical molecule for which there is no known use cannot be patented.

The novelty, utility, and non-obvious requirements, together with the enabling description requirement, work in unison to ensure that only the most deserving innovations receive patent protection and, once a patent is granted, the public is provided with full access to the teachings of the inventors.

In the biotechnology field, the patent review process often takes 2-5 years or more, and typically will involve multiple communications between the applicant and the patent examiner. As a result of the thorough examination given to each application, every granted patent carries with it a presumption of validity (*21*).

To effectively obtain international patent protection, it is important to recognize that, as a general rule, each country has its own patent laws. Patent protection can be obtained in a particular country only if the requirements of that country have been satisfied. Many of the basic requirements for patentability are common to all countries. For example, most patent systems have provisions limiting the availability of patent protection to inventions which are new and involve some significant advance compared to previously known subject matter. Although there are these basic similarities between virtually all patent systems, there are also important differences. For example, there are countries which will not grant patent protection for methods for treating humans (*40-41*). Other countries do not grant patent protection for pharmaceutical or biotechnology inventions (*42*). Therefore, it is possible that an invention may be patentable in one country but not in another.

Virtually every developed country in the world has a patent system designed to foster creativity and expedite the public dissemination of new innovations. Thus, patent systems are not a product of capitalism or any other economic system, nor is the patent system linked to democracy or any other political system. It is even more basic than that—it is simply a means for encouraging creativity and, just as importantly, a means for facilitating the rapid public dissemination of new ideas.

Patents and Court Cases Relating to Patenting of Living Organisms

The patent system has stood the challenges of time in promoting the progress of science and the useful arts in America. The ability to patent a living microbe stands as proof of the responsiveness of our patent system to our expanding, innovative society. The issue of whether a living microbe is patentable was first presented to the United States Patent and Trademark Office (PTO) in the early 1970's. The series of events which led to the Supreme decision recognizing the patentability of living organisms is detailed below.

Bergy's Biologically Pure Culture

An application for a patent was filed on behalf of the applicants, Malcolm E. Bergy, Vedpal Malik and John Coats, in the middle 1970's. The patent application claimed a *process* to make a known and useful antibiotic named lincomycin. The process was a microbiological process using a biologically pure culture of the new microbe named *Streptomyces vellosus*. While the patent application was pending in the PTO, the applicants' attorney added a claim to the "biologically pure culture of *Streptomyces vellosus*" to go along with the process claims already in the patent application. Adding a new claim is not an unusual procedural practice. In fact, patent attorneys frequently amend patent applications by adding, changing or subtracting claims.

The response by the Patent Examiner to this added claim was unequivocal: Not patentable. The Examiner's response was not surprising; despite the fact that hundreds of patents had issued in the previous thirty years with *process* claims reciting the use of a new microbe, there had never been a patent granted on a living organism. The PTO Examiner's rejection was based primarily on the position that the biologically pure culture was a "product of nature." The applicants' attorney countered with affidavits from three experts in microbiology who attested that the biologically pure culture at issue was the product of skilled microbiologists doing their work in the sterile confines of a laboratory. Without "man" and special equipment, the biologically pure culture would not exist. To further distinguish the biologically pure culture from what might be found in nature, it was emphasized that the biologically pure culture possessed the valuable property of being able to produce significant amounts of the useful antibiotic lincomycin, a property which clearly distinguished this entity from anything which might exist in nature. The PTO Examiner maintained his position, which left the applicants with a choice of either dropping the issue, or forging ahead with an appeal to the PTO Board of Appeals. The applicants chose to appeal.

The PTO Board heard the appeal argument on May 20, 1976, and shortly thereafter, handed down its 2:1 decision. The Board majority affirmed the Examiner's rejection, but for a reason different than the Examiner's. It held the biologically pure culture to be unpatentable because the culture was "living." Never mind the fact that the Patent Act does not mention either "living" or "dead." The dissenting Board member wrote an opinion favorable to the appellants. The dissenting opinion, combined with the existing appeal record, was considered by the Applicants to be sufficient to take the case up to the next appeal court, The United States Court of Customs and Patent Appeals (CCPA).

The case was briefed and then argued before the CCPA on March 3, 1977. On October 6, 1977, in a landmark 3:2 decision, the Court reversed the PTO Board. The learned Judge Giles S. Rich, who authored the majority opinion, wrote:

"In short, microorganisms have come to be important tools in the chemical industry, especially the pharmaceutical branch thereof, and when a new and useful tangible industrial tool is invented which is unobvious, so that it complies with the prerequisites to patentability...we do not see any reason to deprive it or its creator or owner of the protection and advantages of the patent system by excluding it from the Section 101 categories of patentable invention on the sole ground that it is alive. It is because it is alive that it is useful. The law unhesitatingly grants patent protection to new, useful, and unobvious chemical compounds and compositions, in which category are to be found the *products* of microbiological processes, for example, vitamin B-12 and adrenalin...and countless other pharmaceuticals. We see no sound reason to refuse patent protection to the microorganisms themselves—a kind of tool used by chemists and chemical manufactures in much the same way as they use chemical elements, compounds, and compositions which are not considered to be alive, notwithstanding their capacities to react and to promote reaction to produce new compounds and compositions by chemical processes in much the same way as do microorganisms. We think it is in the public interest to include microorganism within the terms 'manufacture' and 'composition of matter' in Section 101. In short, we think the fact that microorganisms, as distinguished from chemical compounds, are alive is a distinction without legal significance and that disposes of the board's ground of rejection and the sole reason for refusal of a patent argued by the solicitor.

Chakrabarty's Oil-Eating Microbe

While the Bergy applicants were proceeding before the PTO Examiner and Board, unknown to them was the fact that a patent application, filed on behalf of General Electric and naming Ananda Chakrabarty as an inventor, was following a similar route in the PTO. As a matter of fact, Chakrabarty received virtually the same decision and opinions from the Board as did Bergy, and *before* Bergy *et al.* (*31*) received their PTO Board decision. However, Chakrabarty's attorney paused to brief and argue the "living" issue, whereas, Bergy immediately climbed to the next rung of the appeal ladder. All proceedings before the PTO Examiner and Board are non-public; therefore, only the PTO knew that Bergy and Chakrabarty were following the same route.

Though Chakrabarty was also litigating the issue of patenting a living microbe, his invention was not a "biologically pure culture," but rather a novel microbe which was "genetically engineered." Through the transfer of certain plasmids into a single microbe, this new microbe was then able to digest a broader range of hydrocarbons found in oil. With all the publicity about major oil spills darkening our beaches and threatening wildlife, the utility of the engineered microbe was clear.

Chakrabarty followed Bergy to the CCPA, and, as expected, the CCPA reversed the PTO board, relying on its prior Bergy decision.

The sequence of events which followed is important from a legal standpoint, but discussion of these events is beyond the scope of this paper. Suffice it to say that the CCPA came out with a second landmark decision in 1979 on the issue of patenting a living microbe. This decision (4:1) combined both the Bergy and Chakrabarty cases, and was the longest opinion ever written by the CCPA. The CCPA's decision was taken to the Supreme Court and in a 1980 decision the Supreme Court agreed with the CCPA. Thus, it is now established that a living microbe can be patented.

Although both Bergy and Chakrabarty involved the validity of claims drawn to single-celled microorganisms, the language and reasoning of each opinion gives no suggestion that a multicellular organism should be treated any differently. Chief Justice Burger, speaking for the majority in Chakrabarty, displayed a willingness to construe the range of patentable subject matter very broadly. Quoting the legislative history of the current Patent Act, the Chief Justice stated, "Congress intended statutory subject matter to 'include anything under the sun that is made by man'." The Court also noted that, as a matter of statutory construction, the patent statute's use of broad terms to define patentable subject matter is evidence that 'Congress plainly contemplated that the patent laws would be given wide scope'.

More recently, the United States Commissioner of Patents announced in 1987 that the patent office would grant patents on non-human multicellular living organisms, including animals. The Commissioner's announcement came less than a week after the Patent and Trademark Office (PTO) Board of Patent Appeals held *In re Allen* that polyploid oysters are patentable subject matter. The *Allen* decision and subsequent statement by the Commissioner marked the end of a PTO policy whereby patent claims to multicellular animals were automatically rejected on the grounds that animals were not patentable subject matter under the applicable section of the patent statute (*30*).

The PTO's decision to allow patents on multicellular animals is based predominantly on its interpretation of the landmark decisions in the Bergy and Chakrabarty cases.

The Role of Patents in Fostering Commercialization of New Technologies

The Bergy and Chakrabarty decisions were important points in the evolution of biotechnology patent law, and perhaps more importantly, the infant biotechnology industry. These legal decisions provided a critical spark which propelled the fledgling U.S. biotechnology industry forward. In the nearly thirty years since the Bergy and Chakrabarty decisions, the biotechnology field has rapidly expanded into a multibillion dollar industry employing thousands and producing products which will benefit all of mankind. This rapid growth could not have occurred without the investment of enormous sums of time, effort, and money. It is extremely unlikely that such investment could have occurred without a legal mechanism for providing some limited protection for the fruits of this highly speculative research. The proper application of the patent laws by both the PTO and the Courts have provided the necessary environment for this industry to flourish.

The patent system is important not only to protect research at private companies but also to protect the taxpayers' investment in research at public institutions such as universities. Some have argued that inventions at government laboratories and universities should not be patented and, instead, should be free for the taking. However, a careful analysis of these situations reveals that patents can play a crucial role in the effective commercialization of this technology and the equitable distribution of profits which may result from such commercialization.

Take, for example, the discovery by a government researcher of a new microbe, gene or protein with potential commercial value. Typically, the government agency does not have the expertise or resources to take this invention all the way from the laboratory to the market place. Therefore, the technology must be developed by an outside entity. In order for that outside entity to have a realistic chance of recouping its investment, it is critical to have a limited period of exclusivity for that product. Without any prospects for patent protection, a new technology is far less attractive to a potential licensee.

Patents can also play an important beneficial role in the efficient development of university technologies. It is now commonplace for universities to use patents to protect intellectual property created by researchers. However, prior to the use of patents by universities, it was common practice for big companies, and other private entities, to directly contact researchers who had promising technologies. Often, for the price of a dinner, that company could have immediate and complete access to valuable technology. When that company developed the technology, no compensation was given to the university. Rather, that company would reap a windfall from publicly funded research. By contrast, if the technology is patented by the university, the company will be required to obtain a license for the technology and share its profits with the university. Typically, the funds paid to the university from the licensee are distributed among the inventors, the university department from which the invention came, and the general funds of the university. In this way, the taxpayers' money which originally went towards university research has paid dividends in the development of the technology as well as enhanced funding of the university.

Conclusion

The Patent Office rulings and the appellate court decisions have played a vital role in creating the proper legal environment for the growth and development of the environmental biotechnology industry. This environment provides incentives to rapidly disseminate information through the patent system as well as rewards for those who develop highly innovative and useful technologies. The existence of the proper incentives for growth in this industry is reflected in the number of patents which have issued. For example, the U.S. Patent Office Web Site lists *zero* bioremediation patents from 1976-1985 (Table 1). Thus, the impact on the environmental biotechnology industry of the appellate court decisions recognizing the broad applicability of the patent laws has been extensive and highly beneficial. Our search parameters identified only 11 such patents during this time period (Table 1). However, as the biotechnology revolution gained momentum, the number of these patents has rapidly increased. In 1991 there were 6 bioremediation patents, 8 in 1992, 8 in 1993, 9 in 1994, 26 in 1995, 27 in 1996, and 20 in 1997-98 (Table 1). The Supreme Court's decision to uphold the ideas of Jefferson has led to the establishment of a necessary and vital industry. Whether this industry is termed bioremediation or environmental mitigation, or some other appropriate term, patent protection has played, and will continue to play, an important role in the development of this important industry.

Table I. United States Environmental Mitigation Patents by Year

Patent No.	*Issue Date*	*Title*
3,849,104	1974	Control of northern jointvetch with *Colletotrichum gloeosporioides* penz. *Aeschynomene* isolates of *Pythium* species which are antagonistic to *Pythium ultimum*
3,999,973	1976	Control of prickly sida and other weeds with *Colletotrichum malvarum*
4,162,912	1979	Composition and process for controlling milkweedvine
4,263,036	1981	Method and composition for controlling *Hydrilla*
4,274,955	1981	Process for the degradation of cyanuric acid
4,390,360	1983	Control of sicklepod, showy crotalaria, and coffee senna with *Alternaria cassiae*
4,419,120	1983	Control of prickly sida, velvetleaf, and spurred anoda with fungal pathogens
4,511,657	1985	Treatment of obnoxious chemical wastes
4,521,515	1985	Bacterial strain for purifying hydrocarbons pollution and purification process
4,535,061	1985	Bacteria capable of dissimilation of environmentally persistent chemical compounds
4,554,075	1985	Process of degrading chloro-organics by white-rot fungi
4,556,638	1985	Microorganism capable of degrading phenolics
4,593,003	1986	Bacterial method and compositions for isoprenoid degradation
4,609,550	1986	*Bacillus cereus* subspecies israelensis toxic to Diptera larvae
*4,962,034	1986	Bioremediation of organic contaminated soil and apparatus therefore
*4,850, 745	1986	Bioremediation system
4,643,756	1987	Bioherbicide for Florida beggarweed

Table I. United States Environmental Mitigation Patents by Year

Patent No.	*Issue Date*	*Title*
4,695,462	1987	Cellular encapsulation of biological pesticides
5,000,000	1991	Ethanol production by *Escherichia coli* strains co-expressing *Zymomonas* pdc and adh genes
*5,062,956	1991	Bioremediation of chromium (VI) contaminated aqueous systems by sulfate reducing bacteria
*5,059,252	1991	Method for enhancing bioremediation of hydrocarbon contaminated soils
*5,057,221	1991	Aerobic biological dehalogenation reactor
*5,037,551	1991	High-flow rate capacity aerobic biological dehalogenation reactor
*4,992,174	1991	Fixed bed bioreactor remediation system
*5,160,525	1992	Bioremediation enzymatic composition
*5,160,488	1992	Bioremediation yeast and surfactant composition
*5,158,595	1992	Soil bioremediation enzymatic composition
*5,155,042	1992	Bioremediation of chromium (VI) contaminated soil residues
*5,133,625	1992	Method and apparatus for subsurface bioremediation
*5,132,224	1992	Biological remediation of creosote- and similarly-contaminated sites
*5,100,455	1992	Process of bioremediation of soils
*5,080,782	1992	Apparatus for bioremediation of sites contaminated with harzardous substances
*5,265,674	1993	Enhancement of in situ microbial remediation of aquifers
*5,264,018	1993	Use of metallic peroxides in bioremediation
*5,258,303	1993	Bioremediation system and method
*5,232,596	1993	Bio-slurry reaction system and process for hazardous waste treatment

*Patents from patent office web site: http/www.patents.uspto.gov

Continued on next page.

Table I. United States Environmental Mitigation Patents by Year

Patent No.	*Issue Date*	*Title*
*5,227,069	1993	Bioremediation method
*5,225,083	1993	Method for bioremediation of grease traps
*5,200,080	1993	Waste treatment oxidation operations
*5,178,491	1993	Vapor-phase nutrient delivery system for in situ bioremediation of soil
*5,369,031	1994	Bioremediation of polar organic compounds
*5,364,787	1994	Genes and enzymes involved in the microbial degradation of pentachlorophenol
*5,362,397	1994	Method for the biodegradation of organic contaminants in a mass of particulate solid
*5,342,769	1994	Microbial dehalogenation using methanosarcia
*5,340,376	1994	Controlled-released microbe nutrients and method for bioremediation
*5,336,290	1994	Semi-solid activated sludge bioremediation of hydrocarbon-affected soil
*5,302,286	1994	Method apparatus for in situ groundwater remediation
*5,300,227	1994	Bioremediation of hydrocarbon contaminated soils and water
*5,286,140	1994	Bioremediation systems and methods
*5,478,464	1995	Apparatus for the biodegradation of organic contaminants in a mass of particulate solids
*5,476,992	1995	In situ remediation of contaminated heterogeneous soils
*5,476,788	1995	Solid phase bioremediation methods using lignin-degrading fungi
*5,472,294	1995	Contaminant remediation, biodegradation and volatilization methods and apparatuses

Table I. United States Environmental Mitigation Patents by Year

Patent No.	*Issue Date*	*Title*
*5,470,742	1995	Dehalogenation of organohalogen-containing compounds
*5,466,600	1995	Use of carbon monoxide dehydrogenase for bioremediation of toxic compounds
*5,466,590	1995	Constitutive expression of P450soy and ferredoxin-soy in Streptomyces
*5,464,771	1995	Biologically pure culture of *Actinomyces viscosus* strain used for the bioremediation of chlorinated hydrocarbons
*5,458,747	1995	In situ bio-electrokinetic remediation of contaminated soils containing hazardous mixed wastes
*5,455,173	1995	Biological isolates for degrading nitroaromatics and nitramines in water and soils
*5,443,845	1995	Composition for enhanced bioremediation of petroleum
*5,441,885	1995	Bacterial strains for bioremediation
*5,436,160	1995	Bioremediation of hydrocarbon contaminated soil
*5,431,717	1995	Method for rendering refractory sulfide ores more susceptible to biooxidation
*5,427,944	1995	Bioremediation of polycyclic aromatic hydrocarbon-contaminated soil
*5,414,198	1995	Degradation of nitrocellulose by combined cultures of *Sclerotium rolfsii* atcc 24459 and *Fusarium solani* IFO 31093
*5,413,713	1995	Method for increasing the rate of anaerobic bioremediation in a bioreactor
*5,403,809	1995	Composite inorganic supports containing carbon for bioremediation
*5,403,799	1995	Process upset-resistant inorganic supports for bioremediation

Continued on next page.

Table I. United States Environmental Mitigation Patents by Year

Patent No.	*Issue Date*	*Title*
*5,398,756	1995	In situ remediation of contaminated soils
*5,397,755	1995	Low density glassy materials for bioremediation supports
*5,395,808	1995	Inorganic supports for bioremediation
*5,395,419	1995	Therapeutic and preventative treatment of anaerobic plant and soil conditions
*5,389,248	1995	Bioreactor for biological treatment of contaminated water
*5,387,271	1995	Biological system for degrading nitroaromatics in water and soils
*5,384,048	1995	Bioremediation of contaminated groundwater
*5,587,079	1996	Process for treating solutions containing sulfate and metal ions.
*5,585,272	1996	Solid phase system for aerobic degradation
*5,583,041	1996	Degradation of polyhalogenated biphenyl compounds with white-rot fungus grown on sugar beet pulp
*5,577,558	1996	In-well device for in situ removal of underground contaminants
*5,573,575	1996	Method for rendering refractory sulfide ores more susceptible to biooxidation
*5,571,715	1996	Surface active metal chelated nutrients for bioremediation of hydrocarbon contaminated soils and water
*5,571,705	1996	Biofilter for bioremediation of compound
*5,570,973	1996	Method and system for bioremediation of contaminated soil using inoculated diatomaceous earth
*5,569,634	1996	Process upset-resistant inorganic supports for bioremediation

Table I. United States Environmental Mitigation Patents by Year

Patent No.	*Issue Date*	*Title*
*5,567,324	1996	Method of biodegrading hydrophobic organic compounds
*5,562,588	1996	Process for the in situ bioremediation of Cr (VI)-bearing solids
*5,561,059	1996	Substrate bioavailability enhancing chemical mixture for use in bioremediation
*5,561,056	1996	Class of bifunctional additives for bioremediation of hydrocarbon contaminated soils and water
*5,560,737	1996	Pneumatic fracturing and multicomponent injection enhancement of in situ bioremediation
*5,545,801	1996	Wand inductor for remediation of contaminated soil
*5,531,898	1996	Sewage and contamination remediation and materials for effecting same
*5,525,139	1996	Process for bioremediation of soils
*5,523,217	1996	Fingerprinting bacterial strains using repetitive DNA sequence amplification
*5,518,910	1996	Low density glassy materials for bioremediation supports
*5,514,588	1996	Surfactant-nutrients for bioremediation of hydrocarbon contaminated soils and water
*5,512,478	1996	Genes and enzymes involved in the microbial degradation of pentachlorophenol
*5,508,194	1996	Nutrient medium for the bioremediation of polycyclic aromatic hydrocarbon-contaminated soil
*5,503,774	1996	Class of bifunctional additives for bioremediation of hydrocarbon contaminated soils and water
*5,501,973	1996	Treatment for contaminated material
*5,492,881	1996	Sorbent system

Continued on next page.

Table I. United States Environmental Mitigation Patents by Year

Patent No.	*Issue Date*	*Title*
*5,486,474	1996	Bioremediation method using a high nitrogen-containing culture of white rot fungi on sugar beet pulp
*5,480,549	1996	Method for phosphate-accelerated bioremediation
*5,763,815	1997-98	Apparatus for bioemediating explosives
*5,756,304	1997-98	Screening of microorganisms for bioremediation
*5,753,109	1997-98	Apparatus and method for phosphate-accelerated bioremediation
*5,744,105	1997-98	Slurry reactor
*5,741,427	1997-98	Soil and/or groundwater remediation process
*5,736,669	1997-98	Systems for bioremediating explosives
*5,734,086	1997-98	Cytochrome p 450.sub.lpr gene and its uses
*5,733,067	1997-98	Method and system for bioremediation of contaminated soil using inoculated support spheres
*5,730,550	1997-98	Method for placement of a permeable remediation zone in situ
*5,725,885	1997-98	Composition for enhanced bioremediation of oil
*5,716,839	1997-98	Phytoadditives for enhanced soil bioremediation
*5,705,690	1997-98	Urea-surfactant clathrates and their use in bioremediation of hydrocarbon contaminated soils and water
*5,698,441	1997-98	Surfactant formulations containing menthadiene and menthadiene alcohol mixtures for enhanced soil bioremediation
*5,691,136	1997-98	Fingerprinting bacterial strains using repetitive DNA sequence amplification
*5,690,173	1997-98	Apparatus for enhanced bioremediation of underground contaminants

Table I. United States Environmental Mitigation Patents by Year

Patent No.	*Issue Date*	*Title*
*5,688,685	1997-98	System and methods for biodegradation of compounds
*5,688,304	1997-98	Method for improving the heap biooxidation rate of refractory sulfide ore particles that are biooxidized using recycled bioleachate solution
*5,686,299	1997-98	Method and apparatus for determining nutrient stimulation of biological processes
*5,685,891	1997-98	Composting methods
*5,681,739	1997-98	Method for in situ or ex situ bioremediation of hexavalent chromium contaminated soils and/or groundwater
*5,679,364	1997-98	Compositions and methods for reducing the amount of contaminants in aquatic and terrestrial environments
*5,678,639	1997-98	Self-contained bioremediation unit with dual auger head assembly
*5,668,294	1997-98	Metal resistance sequences and transgenic plants
*5,658,458	1997-98	Apparatus for removing suspended inert solids from a waste stream
*5,656,422	1997-98	Compositions and methods for detection of 2,4-dichlorophenoxyacetic acid and related compounds
*5,656,169	1997-98	Biodegradation process for de-toxifying liquid streams
*5,653,675	1997-98	Bioremediation process for polluted soil water system
*5,653,288	1997-98	Contaminant remediation, biodegradation and volatilization methods and apparatuses
*5,641,679	1997-98	Methods for bioremediation
*5,635,394	1997-98	Arrangement for air purification

Continued on next page.

Table I. United States Environmental Mitigation Patents by Year

Patent No.	*Issue Date*	*Title*
*5,635,392	1997-98	Nutrient mixtures for the bioremediation of polluted soils and waters
*5,633,164	1997-98	Methods for fluid phase biodegradation
*5,627,045	1997-98	Multi-test format with gel-forming matrix for characterization of microorganisms
*5,626,755	1997-98	Method and apparatus for waste digestion using multiple biological processes
*5,626,437	1997-98	Method for in situ bioremediation of contaminated ground water
*5,624,843	1997-98	Nutrient additives for bioremediation of hydrocarbon contaminated waters
*5,618,727	1997-98	Bioremediation process design utilizing in situ soil washing
*5,618,329	1997-98	Bioremediation of hydrocarbon contaminated soils and water
*5,616,162	1997-98	Biological system for degrading nitroaromatics in water and soils
*5,614,410	1997-98	Bioremediation of soil or groundwater contaminated with compounds in creosote by two-stage biodegradation
*5,614,097	1997-98	Compositions and method of use of constructed microbial mats
*5,611,839	1997-98	Method for rendering refractory sulfide ores more susceptible to biooxidation
*5,611,837	1997-98	Bioremediation method
*5,610,065	1997-98	Integrated chemical/biological treatment of organic waste
*5,610,061	1997-98	Microorganisms for biodegrading compounds
*5,609,667	1997-98	Process and material for bioremediation of hydrocarbon contaminated soils

Table I. United States Environmental Mitigation Patents by Year

Patent No.	*Issue Date*	*Title*
*5,595,893	1997-98	Immobilization of microorganisms on a support made of synthetic polymer and plant material
*5,593,888	1997-98	Method for accelerated bioremediation and method of using an apparatus therefore
*5,593,883	1997-98	Ancient microorganisms
*5,591,341	1997-98	Method and system for water bioremediation utilizing a conical attached algal culture system

*Patents from patent office web site: http/www.patents.uspto.gov

Acknowledgments

Florida Agricultural Experiment Station Journal Series No. R-06612.

Literature Cited

1. Alexander, M. *Biodegradation and bioremediation*; Academic Press: San Diego, CA, **1994**; p 302.
2. Alleman, B. C.; Leeson, A., Eds.; *In situ and on-site bioremediation: papers from the Fourth International In Situ and On-Site Bioremediation Symposium, New Orleans, April 28-May 1, 1997*. Battelle Press: Columbus, OH.
3. Baker, K. H.; Herson, D. S., Eds.; *Bioremediation*; McGraw-Hill: NY, **1994**; 375 p.
4. Bajpai, R. K.; Zappi, M. E., Eds.; *Bioremediation of surface and subsurface contamination*; New York Academy of Sciences: NY, **1997**; Vol. 89.
5. Chaudhry, G. R., Ed.; *Biological degradation and bioremediation of toxic chemicals*; Dioscorides Press: Portland, OR, **1994**; 515 p.
6. Clark, N.; Crull, A., Eds.; *Bioremediation of hazardous wastes, wastewater, and municipal waste*; Business Communications Co.: Norwalk, CT, **1997**; 253 p.
7. Cookson, J. T. *Bioremediation engineering: design and application*; McGraw-Hill: NY, **1995**; 524 p.
8. Crawford, R. L.; Crawford, D. L., Eds.; *Bioremediation: principles and applications*; Cambridge University Press: Cambridge, NY, **1996**; 400 p.
9. Eweis, J. B. *Bioremediation principles*; WCB/McGraw-Hill: Boston, MA, **1997**.
10. Glass, D. J. *The promising worldwide bioremediation market*; Decision Resources: Waltham, MA, **1993**; Vol. 8.
11. Hickey, R. F.; Smith, G., Eds.; *Biotechnology in industrial waste treatment and bioremediation*; CRC: Boca Raton, FL, **1996**; 379 p.
12. Hinchee, R. E.; Anderson, D. B.; Hoeppel, R. E. *Bioremediation of recalcitrant organics*; Battelle Press: Columbus, OH, **1995**; 368 p.
13. Hinchee, R. E.; Brockman, F. J.; Vogel, C. M., Eds.; *Microbial process for bioremediation*; Battelle Press: Columbus, OH, **1995**; 361 p.
14. Hinchee, R. E.; Douglas, G. S.; Ong, S. K. *Monitoring and verification of bioremediation*; Battelle Press: Columbus, OH, **1995**; 273 p.
15. Hinchee, R. E.; Kittel, J. A.; Reisinger, H. J. *Applied bioremediation of petroleum hydrocarbons*; Battelle Press: Columbus, OH, **1995**; 534 p.
16. Hinchee, R. E.; Leeson, A.; Semprini, L. *Bioremediation of chlorinated solvents*; Battelle Press: Columbus, OH, **1995**; 338 p.
17. Hinchee, R. E.; Means, J. L.; Burris, D. R. *Bioremediation of inorganics*; Battelle Press: Columbus, OH, **1995**; 174 p.
18. Jennings Group. *U. S. Bioremediation Market, 1994-2000: A comprehensive market and business analysis*; DEVO Enterprises: Washington, DC, **1993**; 182 p.

19. King, R. B.; Long, G. M.; Sheldon. J. K. *Practical environmental bioremediation*; Lewis Publishers: Boca Raton, FL, **1997**.
20. Riser-Roberts, E. *Bioremediation of petroleum contaminated sites*; C. K. Smoley: Chelsea, MI, **1992**.
21. Schepart, B. S., Ed.; *Bioremediation of pollutants in soil and water*; ASTM: Philadelphia, PA, **1995**; 259 p.
22. Skipper, H. D.; Turco, R. F. *Bioremediation: science and application*; Soil Science Society of America: Madison, WI, **1995**; 322 p.
23. U. S. Environmental Protection Agency. *Bioremediation of hazardous wastes: research, development, and field evaluations*; Biosystem Technology Development Program, Office of Research and Development, U.S. EPA: Washington, DC, **1995**; 129 p.
24. U.S. Environmental Protection Agency. *Symposium on bioremediation of hazardous wastes: research, development, and field evaluations;* Abstracts: Rye Town Hilton, Rye Brook NY, August 8-10, 1995; Office of Research and Development: Washington, DC, **1995**; 127 p.
25. Wild, J. R. Varfolomeyer, S. D.; Scozzafava, A., Eds.; *Perspectives in bioremediation: technologies for environmental improvement*; Kluwer Academic Publishers: Dordrecht, Boston, MA, **1997**; 123 p.
26. Young, C. W., Bahn, L. E.; Copley-Graves, L., Eds.; *Bioremediation series commutative indices: 1991-1995*; Battelle Press: Columbus, OH, **1995**; 170 p.
27. U.S. Constitution Article 1 Section 8 clause 8.
28. V Writings of Thomas Jefferson at pages 75-76.
29. *Diamond v. Chakrabarty*, 206 USPQ 193, United States Supreme Court 1980.
30. See, for example, 35 United States Code § 101 (describing subject matter which can be patented) and § 112 (describing the requirement for the patent applicant to provide a full written description of the invention).
31. *In re Bergy*, Court of Customs and Patent Appeals, 201 USPQ 352 (1979).
32. *Diamond v. Chakrabarty*, 206 USPQ 193, United States Supreme Court 1980.
33. 35 USC § 112.
34. 35 USC § 102.
35. 35 USC §112 requires the patent applicant to provide a complete written description of how to make and use the invention. This written description is published for all the world to see when the patent is granted.
36. International patent applications are published 18 months from the original filing date, regardless of whether patent protection is granted.
37. 35 USC § 282.
38. 35 USC § 103.
39. 35 USC § 101.
40. The European Patent Office, for example, does not grant patent protection for methods of treating humans.
41. India, for example, does not grant patents on pharmaceutical inventions.
42. For a comparison of trade secret and patent protection see R. Saliwanchik, "Protecting Biotechnology Inventions: A Guide for Scientists" Science Tech/Springer-Verlag, 1988.

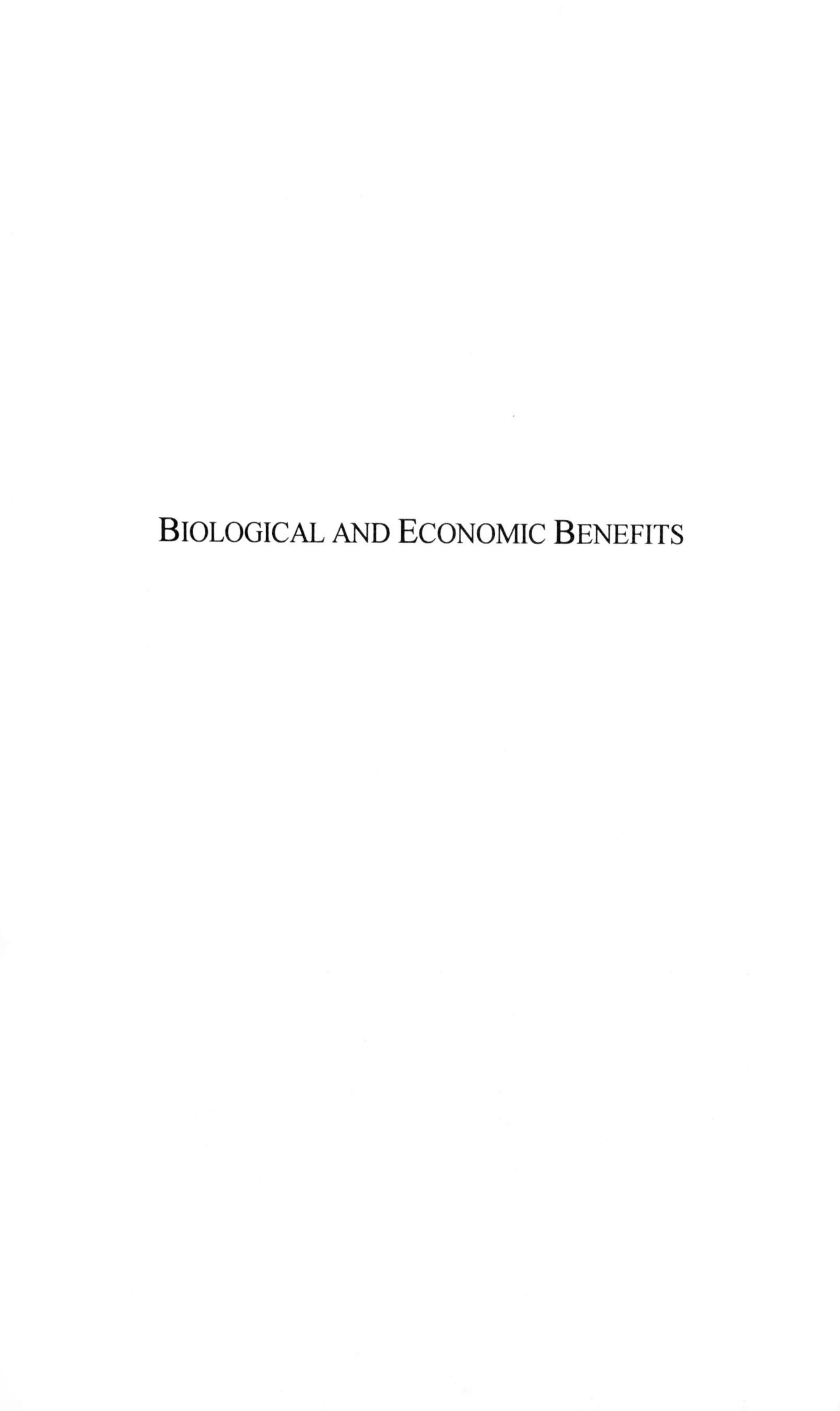

Biological and Economic Benefits

Chapter 11

The Role of Benefits in the Regulatory Arena

Nancy N. Ragsdale [1] and Ronald E. Stinner [2]

[1] **Agricultural Research Service, U. S. Department of Agriculture, Room 331, Building 005, BARC-W, Beltsville, MD 20705-2350**
[2] **Center for Integrated Pest Management, National Science Foundation, 1017 Main Campus Drive, Suite 1100, NCSU Centennial Campus, Raleigh, NC 27606**

Although the decision-making process involved in determining the registration status of pesticides no longer incorporates, in most cases, consideration of benefits from using the compound or material under scrutiny, agriculture has a direct challenge to inform the public of the consequences of proposed regulatory action. Benefits refer to the advantages that may be gained in the yield and/or quality of a treated commodity when compared to the losses incurred when pests are not controlled or when they are managed with materials other than the pesticide under evaluation. In order to assess the benefits of pesticide use, there are four primary areas one must address: information/databases, quality assurance, analysis/output structure, and peer review. Data are needed on pest, pesticide and crop to estimate crop loss and potential net economic benefits from specific pesticide use. Protocols developed in the four primary areas will assure scientifically justified, reproducible assessments that have received review from stakeholder representation.

The goal of agriculture is an available, nutritious, and affordable food and fiber supply. Pest management is one of the essential ingredients needed to achieve this goal. Pesticides play an important role in pest management, and agriculture is frequently called upon to provide information on the benefits of their use. When we refer to benefits in relation to pesticides we mean the advantages gained in agricultural production through the use of these chemicals. This involves an examination of the impacts of changing pest management tactics. It involves such concepts as economic thresholds, injury levels and crop loss. This chapter will primarily focus on the biological data needed for an economic analysis of the impacts resulting from a change in pest control tactics.

Pesticide Use

The use of pesticides in agricultural production is controversial; some say this practice leads to risks that are unacceptable while others indicate the benefits of use far outweigh the risks. To some extent the questions surrounding risks and benefits rise from an uninformed public sector, of which only two percent are directly involved in agricultural production (*1*). Agriculture plays an important role in the United States (U.S.) economy, and the use of sound farming practices to assure abundant production is very important. Preliminary figures indicate that in 1996, agricultural exports totaled 10% of the U.S. export trade (*2*). The U.S. population spent 10.9% of disposable income on food in 1996 (*3*) compared with 20.6% in 1950 (Clauson, A.; Manchester, A., Economic Research Service, USDA, unpublished data). The use of pesticides has undoubtedly contributed to these figures. In 1995 pesticide user purchases in the U.S. equaled approximately one third of the world market in dollars and represented about one fifth of the quantity of active ingredient sold worldwide (*4*). About 7.5 billion dollars is spent per year in the U.S. for agricultural pesticides; herbicides account for about two thirds of that (*5*).

One can gain an idea of the role pesticides play in agricultural production by examining information released by the U.S. Department of Agriculture (USDA) for the 1994 - 95 survey data collected and analyzed by the National Agricultural Statistics Service, USDA (*6*). Table I gives an indication of the role of pesticides in the production of selected crops. These figures do not include seed treatment or post harvest treatments.

Table I. Percent of Acres Receiving Field Applications of Pesticides in Major Producing States for the Respective Commodities (6).

Crop	*Herbicide*	*Insecticide*	*Fungicide*
Corn	97	27	—
Upland Cotton	96	73	9
Fall Potatoes	85	88	83
Soybeans	98	2	—
Winter Wheat	53	8	1
Apples	63	98	93
Oranges	97	94	69
Peaches	66	97	97
Grapes	74	67	90
Strawberries	41	88	89
Raspberries	92	83	90
Fresh Cabbage	55	97	60
Broccoli	67	96	36
Carrots	72	34	71
Fresh Tomatoes	52	94	91
Fresh Snap Beans	60	79	63
Fresh Spinach	52	75	46

Farmers use pesticides to increase yields and to substitute for labor, machinery and fuel.

Agrochemical technology plays an important role in meeting world nutritional needs, achieving sustained development and establishing a quality environment (7).

Legislative History

The Federal Insecticide, Fungicide, and Rodenticide Act (FIFRA) is one of the two primary statutes by which the U.S. Environmental Protection Agency (EPA) regulates pesticides, the other being the Federal Food, Drug, and Cosmetic Act (FFDCA). In 1975 Congress amended FIFRA, requiring that EPA consider the impact of adverse regulatory actions on the production and prices of agricultural commodities. In addition, the 1975 amendment required EPA to notify the Secretary of Agriculture regarding proposals to change or cancel the registration status of a pesticide, giving agriculture an opportunity to state the benefits of pesticide use. This amendment also directed EPA to consider the impact on agriculture before initiating pesticide cancellation proceedings. The consideration of benefits in the regulatory process is also included in FIFRA Section 2 (bb), which directs EPA to examine any unreasonable risk to man or the environment, taking into account the economic, social, and environmental costs and benefits of the use of any pesticide. Thus, the impacts of inadequate pest control through the loss of pest management chemicals became part of pesticide regulation. These impacts were usually described from an economic perspective, expressing losses to producers and changes in costs to consumers in monetary terms. Since maintaining registered uses of a pesticide would negate the impacts and thus benefit producers and consumers, the process was called benefits assessment.

The Food Quality Protection Act (FQPA) of 1996 amended FIFRA and FFDCA so that the direct role of benefits in the regulatory process was considerably diminished (*8*). Benefits of pesticide use will no longer be considered in establishing new tolerances and can only be used to maintain existing tolerances in situations involving pesticides that have been classified as carcinogens. A tolerance for such a pesticide may be retained if its use protects consumers from adverse health effects posing a greater risk than the pesticide or if use of the pesticide is important in avoiding a significant disruption in the domestic supply of an adequate, wholesome, and economical food supply. The chance of using benefits in such a situation to justify retention of a pesticide that has been classified as a carcinogen appears highly unlikely since public reaction would undoubtedly be quite negative. However, determining the benefits of pesticides in agricultural production is still very important. This process offers a mechanism to select efficient pest management systems. EPA will also use such information when comparing pesticides in risk mitigation decisions. In addition, the agricultural community must be in the position to indicate the ramifications of losing pest management tools.

USDA Activities in Providing Benefits Information

During the period immediately preceding passage of the 1975 FIFRA amendment, the USDA and their partners in the states became increasingly aware of the need for an organized and coordinated effort to effectively respond to the regulatory activity of EPA. After this amendment passed, the USDA and representatives from the State Agricultural Experiment Stations as well as the Cooperative Extension Service developed a plan, and the National Agricultural Pesticide Impact Assessment Program (NAPIAP), with federal and state components, was implemented in 1976. The primary mission of NAPIAP was the coordination of activities to promote informed regulatory decisions on pesticides that significantly benefit U.S. agriculture without causing unreasonable adverse effects to human health or the environment. A chief function to achieve this mission was to provide benefits information for use in the regulatory process.

In order to assess the benefits for a specific pesticide, a group of pesticides, or for pest management in a particular commodity, a data gathering process occurred, followed by a biological assessment. This biological assessment, which was based on current approaches to pest control compared to approaches that would be used if a specified pesticide(s) were no longer available, required the following data:

1. Use data--crop being analyzed, acres grown, acres treated for pests, pest(s) of economic importance, pest control agents (chemical and non-chemical), rate applied, method of application, and number of applications.
2. Use-associated factors--crop production cost, crop yield, crop quality, crop production price, and use of the chemical in pest resistance management as well as integrated pest management (IPM).
3. Biological impact resulting from removal of specified pest control methods (such as a specific pesticide)--identification of alternative pesticides (only those registered at the time were considered) and/or practices; and yield, quality (price deduction), and cost changes that occur when using alternative practices.
4. Economic impact analysis--based on the information provided in items 1-3 with net economic effects based on Marchallian demand-and-supply curves (*9*).

There were problems in this assessment process from the very beginning. The most elusive data were, and continue to be, figures pertaining to crop yield and quality changes under different pest management regimes. There is tremendous variation nationwide for any one crop-pest situation. Approaches vary among the disciplines of weed science (*10*), entomology (*11*), and plant pathology (*12*). Climatic differences between years are but one example of the variations that must be taken into consideration. Although models may be developed, they are restrained to specific conditions, and changes in the production system can invalidate the model (*13*). As a result, the crop yield and quality data (often called crop loss) invariably involve the use of estimates based on the scientific knowledge and experience of professionals associated with the particular crop and pest discipline in question. The lack of guidelines and standard procedures, particularly pertaining to crop loss estimates, opened the assessment process to criticism. When NAPIAP was established, the problems associated with a lack of crop loss data were recognized, and the Cooperative State Research Service of the USDA formed a National Crop Loss Design Committee. This committee functioned for several years, but never developed a standard methodology that could be generally accepted, so the committee disbanded. Since state scientists are frequently called upon to provide expert data on crop losses to support the use of pest management tactics, some individuals continued gathering information related to losses (*14*). Other efforts to estimate losses have been made in the disciplines of weed science (*15*) and entomology (*16*), but these have not been incorporated into a nationally accepted standard for benefits methodology.

In contrast to benefits methodology, EPA has developed extensive protocols for the components of risk. After procedures are developed, extensive scientific review and a public comment period occur. The assessment of risks that are actually quite complicated if approached epidemiologically (actual occurrence), is organized into an approach that is transparent, can be examined step by step, and repeated or recalculated if necessary. For example, cancer risk is not usually assessed from an epidemiology study of a human population. Most estimations are derived from the results of laboratory studies using animal species, followed by the use of models to extrapolate the cancer risk. These results may not represent what actually occurs in a human population at measured or estimated exposure levels, but they are a tool that has been used effectively in the regulatory process. Unfortunately, these model-based estimations have been used by the press to present somewhat erroneous impressions of actual risks occurring from the use of pesticides.

Scientists associated with the pest disciplines and commodity production have indicated that laboratory studies or small plot experiments can, in no way, provide an accurate forecast of crop losses that result from inadequate pest management. Thus NAPIAP used what is essentially the epidemiological approach in gathering information to compare crop losses under various pest management regimes. Due to the lack of actual data on losses, scientists have used their expertise to estimate what average losses would result. In February, 1995, a review of NAPIAP indicated that a formal procedure should be developed for establishing assessment-specific protocols. The review panel expressed concern about the benefits assessments that had been generated to date, pointing out lack of documentation and methodology that would permit reproducing results. The panel recommended that data should be collected, analyzed, and reported in a transparent, scientifically rigorous and documented manner such that conclusions could be substantiated and/or reproduced.

Formation of the Benefits Methodology Working Group

To refine the benefits assessment process, a Benefits Assessment Protocols Working Group, with representation from USDA, EPA and the American Crop Protection Association (ACPA), was formed in 1995. The National Science Foundation (NSF) Center for Integrated Pest Management (IPM), North Carolina State University, provided coordination. This group discussed a wide variety of benefits resulting from pest control including human health, quality of life, direct financial returns to producers and consumer advantages resulting from affordable food and fiber. In addition, benefits were recognized that result from the role pesticides play in IPM and from access to a variety of pesticides to manage pest resistance. An agreement was reached that initial methodology efforts would focus on direct financial gains or losses to agricultural producers and consumers. In order to determine financial impacts, biological data must be available to indicate the impact of pests on yield/quality, pest distribution, and the relationship between controlling pests and damage levels. Models could be used to estimate changes in yield and quality, but scientific expertise would remain a factor in interpreting the data (*17*).

Workshop on Benefits Assessment Protocols

The NSF Center for IPM hosted a workshop, sponsored by USDA and EPA, in October, 1997, on developing benefits assessment protocols (18). The invited participants included university scientists and extension specialists, as well as individuals providing perspectives from USDA, EPA, the agricultural chemical industry, and commodity groups. Workshop participants divided into three discussion groups on weeds, insects and plant pathogens. The groups were charged with identifying the critical parameters affecting yield variability and explaining which of those parameters would be expected to have a geographic component. They were also asked how the quantifiable impacts could be compartmentalized and identified to make the benefits assessment process more transparent (accountable and repeatable).

All groups agreed that information and analyses should be conducted by growing region due to differences in climate, soil types, pest species/subspecies, varietal preferences by the local markets/growers, and divergent cultural perspectives with yield data as the critical factor. Yield parameters need to include the weighted average of mean yields for five years. The variables required to determine the impact of the three general pest types discussed may differ for pest type. Standardization of experiments to determine yields does not appear feasible. Certain of the journals in which such data are reported require relatively standard formats, but not standardized experimental procedures. Thus, the issue of protocols for determining benefits is critical for quality assurance.

In order to compare various pest management materials or approaches, certain types of information are required: profiles of alternatives, status of alternatives (registration, rates, costs, etc.), evaluation of comparative performance, and yield/quality data. In addition, consideration should be given to the value of a material that can be used in resistance management and in IPM.

The workshop participants recommended that an *ad hoc* group, responsible to USDA and EPA, be established to prepare recommendations for specific protocols, designed to estimate pest damage (crop loss), based on the guidelines developed during this workshop. These guidelines encompassed four focus areas that are components of assessments that address the benefits of pest management.

1. Information/Databases.
    a. An inventory, publicly accessible, of scientists by state, commodity, discipline, and other areas of expertise.
    b. A pest database that includes crop loss data/models, resistance information, and geographic/seasonal incidence.
    c. A pesticide database that includes a list of approved or preferred information sources including specific journals and databases (by crop), registered pesticide alternatives, non-chemical alternatives, efficacy, and economic information (market share, usage, price).
    d. Crop statistics that include, at the least, acres planted, distribution, price, and quality measurements (e.g., oil content, fresh market versus canning quality, etc.).

The data from b -d are necessary to estimate crop loss and the potential economic benefits from specific pest management materials or methods. Sources of information used in making assessments must be identified giving authors, methodologies used, and quality. Priority should be given to information from published, peer-reviewed databases/documents, followed by surveys/questionnaires, unpublished experimental data, and last of all, scientific estimates. Information based on this approach can be scientifically justified, while still allowing flexibility where experimental data are lacking, contain divergent results, or are of disputed value.

2. Quality Assurance. The data used in estimating benefits/impacts has been a major

issue in the benefits assessment process. As discussed in the section on information and databases, there are various types of data that can be used to estimate impacts (crop loss). Indices should be established that aid in indicating the breadth and depth of the particular data that is used. For example, in examining experimental data, Dr. Paul Borth of Dow AgroSciences suggested an index that shows various parameters in the research that indicate levels of reliability and quality/value (Table II).

Table II. The Borth Index, Proposed by Paul Borth, Dow AgroSciences as a Mechanism to Describe Experimental Data.

Reliability	Index Description	*In situ* Replication	Geographical Replication	Seasonal (years) Replication	Quality/ Value
HIGH ↑	**Category I**				HIGH ↑
	Repeatable	X	X	X	
	Category II				
	May be	X	--	X	
	Repeatable	--	X	X	
	Category III				
	Circumstantial	--	X	--	
	(demonstrations, etc.)	--	--	X	
	Category IV				
↓ LOW	Not replicated	--	--	--	↓ LOW

Use of the Borth Index would permit a weighting scheme in analyzing experimental data. Data and databases which have higher levels of replication would be weighted more heavily in reaching a conclusion. A similar approach could be used for other types of data such as surveys and questionnaires. Protocols to assure the quality of data are a necessity for accurate and reproducible benefits assessments.

3. Analysis/Output Structure. Standardization in this area can be more readily achieved than in field experiments. Documentation of any analysis or modeling must be addressed. The ability to reproduce the output is critical. Protocols in this focus area should also address output shells or templates, that is, how the data are presented for defined user groups. For example, if the output will be used by the EPA, it should be in a form with direct utility to the regulatory decision process. These protocols can also be designed to provide a better understanding of the results for non-scientists.

4. Peer Review. Protocols are needed to establish who reviews the final reports and how the review process should function. An established process will protect the interests of all affected parties. This review process should include representation from all the stakeholders. For example, if the report assesses the benefits of a specific pesticide, the review process should involve representation from registrants, state agricultural extension services, commodity groups, regulatory agencies, and environmental/safety concerns. Time constraints, conflict resolution and feedback to/from reviewers must be part of the process in this area. All data and analyses have subjective elements, making peer review a prerequisite to release of any benefits assessments.

Conclusions

Pesticides play an important role in U.S. agriculture, but the agricultural community has not taken the necessary steps to credibly present to the public and regulatory community information that accurately reflects the benefits derived from pesticide use. Although current legislation has restricted the use of benefit data in regulatory decision making, the information is still critical from the perspectives of public policy and the need to establish efficient crop production systems. Efforts have been initiated to develop widely accepted methodology for estimating benefits. These efforts should continue with development of specific protocols designed to estimate pest damage.

Literature Cited

1. Ragsdale, N. N.; Sisler, H. D. In *Annual Review of Phytopathology;* Cook, R. J.; Zentmyer, G. A.: Shaner, G., Eds.; Annual Reviews Inc.: Palo Alto, CA, 1994, Vol. 32; pp. 545-557.
2. *Agricultural Statistics 1998.* National Agricultural Statistics Service, USDA, U.S.Government Printing Office: Washington, DC, 1998; p. xv-2.
3. Clauson, A.; Manchester, A. *FoodReview* **1997**, *20, issue 3*, 25-27.
4. Aspelin, A. L. *Pesticides industry Sales and Usage-1994 and 1995 Market Estimates;* U.S. Environmental Protection Agency, Office of Prevention, Pesticides and Toxic Substances: Washington, DC, 1997; 35 pp.
5. *Agricultural Resources and Environmental Indicators*; Anderson, M.; Magleby, R., Eds.; Agricultural Handbook No. 712; Economic Research Service, USDA: Washington, DC, 1997; pp. 116-134.
6. *Agricultural Statistics 1997;* National Agricultural Statistics Service, USDA; U.S. Government Printing Office: Washington, DC 1997; pp xiv-1-xiv-8.
7. Klassen, K. In *Eighth International Congress of Pesticide Chemistry - Options 2000;* Ragsdale, N. N.; Kearney, P. C.; Plimmer, J. R., Eds.; American Chemical Society: Washington, DC, 1995; pp. 1-32.
8. Mintzer, E. S.; Osteen, C. *FoodReview* **1997**, *20, issue 1*, 18-26.
9. Osteen, C.; Esworthy, R. In *Proceedings of the Third National IPM Symposium/Workshop*; Lynch, S.; Greene, C.; Kramer-LeBlanc, C., Eds.; Misc. Pub. No. 1542; USDA: Washington, DC, 1997; pp. 134-135.
10. Mortensen, D. A.; Coble, H. D. *In Economic Thresholds for Integrated Pest Management*; Higley, L. G.; Pedigo, L. P., Eds.; University of Nebraska Press: Lincoln, NE, 1996; pp. 89-113.
11. Hutchins, S. H.; Higley, L. G.; Pedigo, L. P. *J. Econ. Entomol.* **1988**, *81*, 1-8.
12. Backman, P.A.; Jacobi, C. In *Economic Thresholds for Integrated Pest Management*; Higley, L. G.; Pedigo, L. P. Eds.; University of Nebraska Press: Lincoln, NE, 1996; pp. 114-127.
13. Gaunt, R. E. In *Annual Review of Phytopathology*; Webster, R. K.; Zentmyer, G. A.; Shaner, G., Eds.; Annual Reviews Inc.: Palo Alto, CA, 1995; Vol. 33; pp. 119-144.
14. Main, C. E.; Nusser, S. M.; Essex, B. W.; Gurtz, S. K. *Crop Losses in North Carolina due to Plant Disease and Nematodes*; CLAS Systems Documentation, Version 2.2; Plant Pathology Special Publication No.10; North Carolina State University: Raleigh, NC, 1990; 169 pp.

15. *Crop Losses Due to Weeds in the United States - 1992*; Bridges, D. C., Ed. Weed Science Society of America: Champaign, IL, 1992; 403 pp.
16. Schwartz, P. H.; Klassen, W. In *Handbook of Pest Management in Agriculture*; Pimental,D., Ed.; CRC Press: Boca Raton, FL, 1981, Vol. I; pp. 15-77.
17. Stinner, R. In *Proceedings of the Third National IPM Symposium/Workshop*; Lynch, S.; Greene, C.; Kramer-LeBlanc, C., Eds.; Misc. Pub. No. 1542; USDA: Washington, DC, 1997; pp. 132-133.
18. *Proceedings of the Workshop on Benefits Assessment Protocols*; Stinner, R. E., Ed.; CIPM Technical Bulletin 100; NSF Center for Integrated Pest Management, North Carolina State University: Raleigh, NC 16 p.

Chapter 12

Pesticides and Human Health: The Influence of Pesticides on Levels of Naturally-Occurring Plant and Fungal Toxins

Carl K. Winter

Department of Food Science and Technology, University of California, Davis, CA 95616

While the potential health risks from pesticide residues generate significant public, legislative, and regulatory concern, it is possible that agricultural pesticide use may, on occasion, influence dietary risks. It has been proposed that pesticide use may reduce the risks associated with naturally-occurring toxins of plants and fungi by reducing the pest pressures which may stress plants into producing their own toxins or by controlling the fungi responsible for mycotoxin production. Very little direct research has been published investigating such pesticide/natural toxin relationships, however; the limited results have indicated that pesticide use may increase or decrease naturally-occurring toxin levels. Current regulatory programs to examine such relationships are burdened by statutory limitations and jurisdictional issues. Present U.S. pesticide regulations allow only very limited consideration of benefits such as decreased risks from naturally-occurring toxins, and separate federal agencies control the regulation of pesticides and the regulation of naturally-occurring food toxins.

Pesticide residues and their potential human health effects continue to receive considerable public, legislative, and regulatory attention while media accounts of this controversial topic remain frequent. The passage of the Food Quality Protection Act (FQPA) of 1996 (*1*) has presented tremendous challenges to pesticide regulators as they strive to more effectively perform risk assessments that consider factors such as cumulative and aggregate exposure and the special exposure and susceptibility issues of sub-populations (i.e. infants and children). Enforcement of FQPA may result in a significant number of regulatory actions limiting the uses of many pesticides, particularly those which belong to families of chemicals that share common toxicological mechanisms of action such as the organophosphate insecticides, the carbamate insecticides, and the triazine herbicides.

Naturally Occurring Toxins in Food

It has been commonly argued that the level of public concern and regulatory scrutiny directed towards pesticide residues in foods is unwarranted given the low relative risks of pesticide residues in the diet and that such attention detracts from the far more serious food safety concerns of microbiological contamination and nutritional imbalance (*2*). In addition, the National Research Council (NRC) recently concluded that the naturally-occurring components of the diet may present greater theoretical cancer risks than synthetic chemicals such as pesticides (*3*). The finding of the NRC supports more than a decade of research by Ames and coworkers who, by comparing the risks of naturally-occurring and synthetic carcinogens quantitatively using their HERP (Human Exposure/Rodent Potency) index, reached a similar conclusion (*4,5*).

Communicating Risks from Naturally Occurring Toxins. If one accepts the conclusion that naturally-occurring chemicals in food pose much greater potential human health risks than pesticide residues, it is tempting to use this as evidence that pesticides may be receiving excessive regulatory scrutiny and that public concern over pesticide residues is misguided. It is important to realize, however, that public and personal decisions also involve critical elements concerning public values and the acceptability of different types of risks. Risk acceptability, while influenced by the magnitude of the risk calculated in the risk assessment process, also includes a number of qualitative factors such as the voluntariness of exposure, controllability, familiarity, origin, memorability, fairness, effects on children, and the level of trust in institutions (*6*). The direct comparison of seemingly unrelated risks such as pesticide residues and naturally-occurring toxins ignores many of these qualitative factors. This type of risk comparison, according to Covello et al. (*7*), represents a poor risk communication
strategy that is likely to be ineffective and, in fact, may not only fail but also may provoke outrage.

Relationship Between Pesticides and Naturally Occurring Toxins. On closer examination, it appears that the risks of pesticide residues and naturally-occurring toxins may not be completely unrelated. The use of pesticides may, in some cases, actually affect levels of naturally-occurring toxins. This relationship enables the use of a higher ranking comparison, according to Covello et al. (*7*), involving the comparison of the risk of doing something (using pesticides) with the risk of not doing something (unaltered naturally-occurring toxin risk); such a comparison is far more likely to be successful as a risk communication tool than comparing seemingly unrelated risks.

One link between pesticide use and levels of naturally-occurring toxins is based upon the premise that plants, when under stress, may produce their own natural toxins, known as phytoalexins (phyton = plant; alexin = defend) (*8*). By reducing plant stress from insect attack, weed competition, or plant pathogens, pesticides may affect changes in phytoalexin synthesis. Hundreds of different phytoalexins have been identified and their occurrence is comprehensively reviewed by Beier and Nigg (*9*). A variety of stimuli have been shown to induce phytoalexin synthesis such as ultraviolet light and heavy-metal salts (*10*), attack by nematodes (*11*), and viral infections (*12*).

Natural Toxin Examples. Notable phytoalexins include the glycoalkaloids produced from potatoes and the linear furanocoumarins produced by umbelliferous plants such as celery. The major glycoalkaloid from potatoes is α-solanine, while others produced include α-chaconine and leptine I (Figure 1); all of these compounds are inhibitors of cholinesterase enzymes that are presumably synthesized to provide insect resistance. Glycoalkaloid levels have been shown to increase as a result of exposure to light or when potatoes are wounded. Breeding programs to confer insect resistance have led to the development of varieties with high α-solanine content of acute toxicity concern to humans (*9*).

Linear furanocoumarins are notorious for their ability to cause contact dermatitis in field workers handling celery plants and have been shown to intercalate into DNA and RNA. These compounds, which include psoralen, bergapten, isopimpinellin, and xanthotoxin (Figure 2), are photosentizing agents used medicinally to treat skin depigmentation and psoriasis. Animal and human epidemiological studies indicate potential carcinogenic risks from psoralen exposure. Under conditions of stress such as fungal attack (*13*), metal ions (*10*), and acidic fog (*14*), celery plants have produced elevated levels of furanocoumarins. Celery plants bred for pest resistance showed linear furanocoumarin levels elevated from 10- to 15-fold and caused photophytodermatitis in grocery store workers (*9*).

Another mechanism by which pesticide use may influence naturally-occurring toxins is through interactions with mycotoxin-producing fungi that colonize food crops. It seems reasonable that pesticides such as fungicides may interfere with mycotoxin synthesis and could therefore reduce the potential health risks associated with consumption of mycotoxins in food.

The best known and studied mycotoxins are the aflatoxins; these mycotoxins are frequently found in a variety of food products including corn and peanuts (*15*). They are produced by *Aspergillus flavus* and *Aspergillus parasiticus* and have been shown to be potent mutagens, carcinogens, and teratogens. Epidemiological studies indicate that aflatoxins may play a role in the development of human primary hepatocellular carcinoma, either independently or in combination with hepatitis B virus (*9*).

Considerable contemporary toxicological concern also surrounds the fumonisins which are mycotoxins produced by the corn pathogens *Fusarium moniliforme* and *Fusarium proliferatum*. Fumonisin B_1 (Figure 3) was discovered in 1988 by a South African research group investigating the cause of human esophageal cancer in parts of southern Africa (*16*) and fumonisin contamination of corn and corn-based food has since been associated by epidemiological data to high occurrences of esophageal cancer risk in Transkei, South Africa (*17*). Fumonisin B_1 has been shown to be hepatocarcinogenic and hepatotoxic in rats, causes

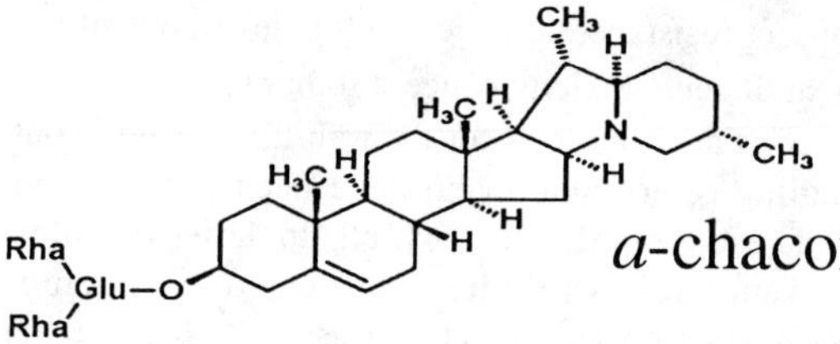

a-chaconine

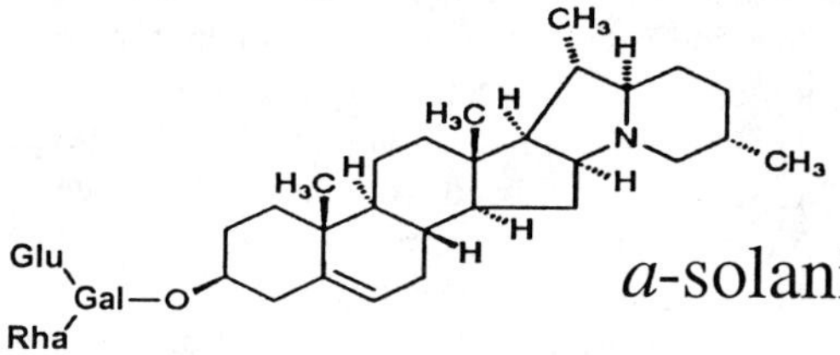

a-solanine

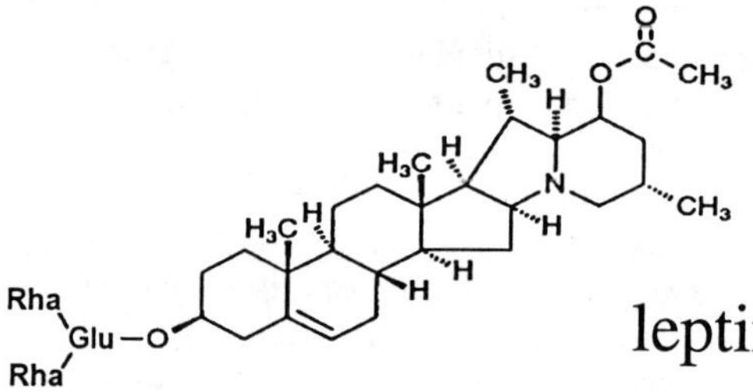

leptine I

Figure1. Potato glycoalkaloids

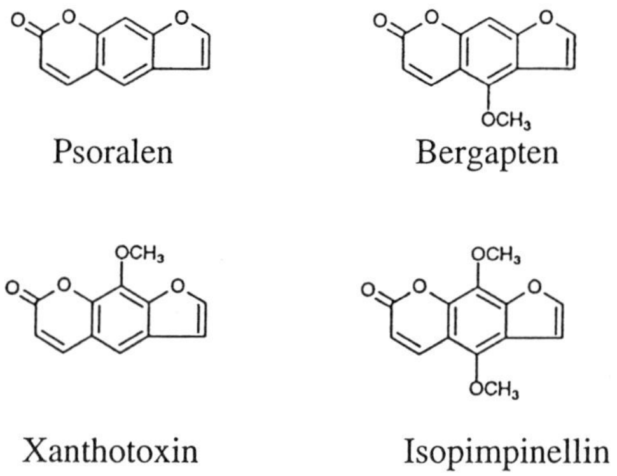

Psoralen

Bergapten

Xanthotoxin

Isopimpinellin

Figure 2. Linear furanocoumarins found in food plants

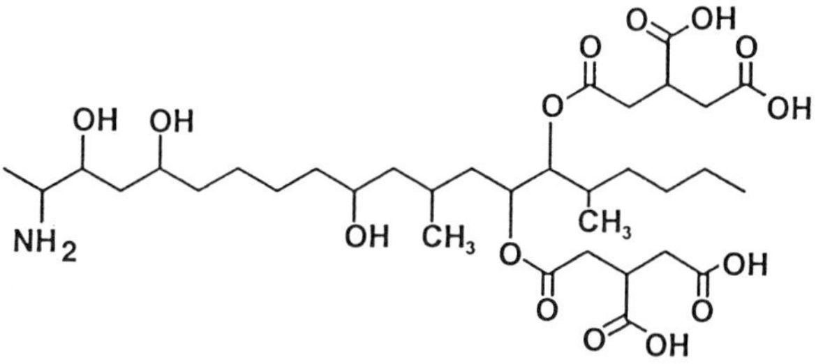

Figure 3. Fumonisin B_1

leukoencephalomalacia in horses, and causes pulmonary edema in pigs. A number of congeners of AAL toxins, which are structurally-related to the fumonisins, have been identified and are produced by the fungus *Alternaria alternata* f.sp. *lycopersici* (*18, 19*) while other toxins such as alternariol, alternariol monomethyl ether, and tenuazonic acid have also been produced by *Alternaria* species (*20*).

Reductions in Naturally Occurring Toxin Levels from Pesticide Use

While it seems plausible that the use of pesticides may reduce levels of phytoalexins and mycotoxins in foodstuffs by reducing plant stress and/or controlling toxin-producing fungi, the available research base investigating such pesticide/natural-toxin relationships is quite sparse, particularly in the case of pesticide/phytoalexin relationships.

Effects of Fungicides. A handful of research papers have been published that investigate the direct relationship between fungicide use and mycotoxin production. Results from fungal culture studies indicate that aflatoxin B_1 levels from *Aspergillus flavus* were reduced by chlorothalonil, dichloran, and mancozeb with chlorothalonil being significantly more effective than the other two fungicides (*21*). Carboxin/captan, tolclofos-methyl/thiram, and procymidone fungicide applications to liquid cultures of *Aspergillus flavus* and to corn grains and sunflower seeds all showed at least some decrease in aflatoxin production at the levels tested (*22*). Iprodione inhibited aflatoxin production from a strain of *Aspergillus parasiticus* grown in culture (*23*), while the use of propionic acid (as ammonium propionate) sprayed on moist unshelled peanuts effectively reduced aflatoxin levels (*24*). The fungicide cuprosan (a mixture of manganese and zinc ethylenebisdithiocarbamates and copper oxychloride), when applied to two strains of *Alternaria alternata* isolated from decayed fruits, inhibited synthesis of alternariol and its monomethyl ether (*25*).

Effects of Insecticides/Nematicides. A small number of studies have investigated the relationship between insecticide/nematicide use and mycotoxin production. While the insecticides may not affect toxigenic fungi directly, they may control damage to food crops that provides opportunities for fungal colonization (*26,27*). Application of the nematicides fenamiphos, carbofuran, and aldicarb reduced the occurrence of *Fusarium* species naturally contaminating roots and fruits of tomato plants and inhibited or reduced production of the mycotoxin zearalenone (*28*). Addition of dichlorvos to culture media of two strains of *Alternaria alternata* and application of dichlorvos to sunflower seeds showed marked decreases in alternariol, alternariol monomethyl ether, and tenuazonic acid levels (*20*). In a different study, sumi oil also reduced levels of alternariol and alternariol monomethyl ether when added to cultures of two strains of *Alternaria alternata* isolated from decayed fruits (*25*).

Increases in Naturally Occurring Toxin Levels from Pesticide Use

Interestingly, some published reports indicate increases in naturally-occurring toxin levels following pesticide application. Applications of a mixture of the fungicides tebuconazole and triadimenol decreased the incidence of *Fusarium* headblight from *Fusarium culmorum* in winter wheat but produced much higher levels of the mycotoxin nivalenol (*29*), indicating that the fungus may itself respond to stress by producing greater levels of toxins. While Trumble et al. (*30*) reported relatively little effect of the herbicide prometryn and the insecticides *Bacillus thuringiensis*, naled, and methomyl on the induction of linear furanocoumarin production in celery, Nigg, et al. (*31*) demonstrated that treatment of a commercial Florida celery cultivar with the fungicides chlorothalonil, manganese ethylenebisdithiocarbamate, and copper hydroxide did not increase psoralen levels but did increase bergapten levels in leaves and stalk by factors of 2 to 4, xanthotoxin levels in stalk by factors of 2 to 3, and isopimpinellin levels in leaves by factors of 2 to 3. Application of the diphenyl ether herbicide acifluorfen to a variety of plants greatly increased the synthesis of several phytoalexins in broad beans (glyceollins, glyceofuran, medicarpin, and wyerone), beans and pinto beans (phaseollin), peas (pisatin), celery (xanthotoxin), and cotton (hemigossypol) (*32*). Such enhancements of natural toxin levels may have been caused by induction of phenylalanine ammonia-lyase, a key enzyme in the synthesis of several phytoalexins (*32*).

Regulatory Implications

Statutory Issues. The potential influence of pesticide applications on levels of naturally-occurring toxins has not received much regulatory attention. The Food Quality Protection Act of 1996 does include provisions that would allow the use of a pesticide that does not meet "reasonable certainty of no harm" criteria in cases where the "use of the pesticide chemical that produces the residue protects consumers from adverse effects on health that would pose a greater risk than the dietary risk from the residue" (*1*). The consideration of such benefits, however, are limited only to non-threshold (cancer) endpoints that present an annual risk of no more than ten times the yearly allowable (1×10^{-6}) risk and no more than two times the lifetime risk (*1*). As such, benefits would not be allowed in cases where the pesticide residue risks exceeded these levels even though the pesticides might reduce the cancer risks from naturally-occurring toxins to a far greater amount.

Practical Issues. Aside from the quantitative statutory restrictions on benefits consideration, practical barriers also exist. The FQPA contains provisions requiring the EPA to publish information concerning the risks and benefits of pesticides to be provided for distribution to consumers at the retail grocery level (*1*). In cases where benefits considerations are used to allow registrations of specific pesticides on particular commodities, it is required that these pesticide/commodity combinations be listed. From a practical standpoint, significant public concern and avoidance of "identified" commodities might be anticipated that would discourage pesticide manufacturers and growers to pursue pesticide registrations on the basis of benefits.

Jurisdictional Issues. The regulatory system currently does not include considerations of cases where the use of pesticides may increase the production of naturally-occurring toxins although the U.S. Food and Drug Administration (FDA) and EPA do consider increases in naturally-occurring toxins from the use of recombinant DNA technologies. This presents an interesting jurisdictional debate as EPA regulates pesticides while FDA has regulatory authority for naturally-occurring toxins.

Conclusions

In summary, it is concluded that knowledge about the potential effects of pesticide use on the production of naturally-occurring toxins is extremely limited. In some cases, pesticide use may decrease levels of natural toxins; in others, natural toxin levels may increase. The human health significance of pesticide/naturally-occurring toxin relationships is also poorly understood; while the NRC concluded that the risks from naturally-occurring toxins may present greater potential human cancer risks than synthetic chemicals such as pesticides in the diet, it was also concluded that most naturally-occurring and synthetic chemicals in the diet appear to be present at levels so low that they are unlikely to pose an appreciable cancer risk or other significant adverse biological effects (*3*). Finally, in the event where the health significance of increasing or decreasing naturally-occurring toxins through pesticide use is deemed important, our present regulatory system is burdened by regulatory statutes and jurisdictional issues that may render regulatory efforts ineffective in terms of protecting public health.

Literature Cited

1. *Food Quality Protection Act of 1996.* Public Law 104-170, 104th Congress,Washington, D.C. 1996.
2. Winter, C.K. *Weed Technol.* **1996**, *10*, 969-973.
3. *Carcinogens and Anticarcinogens in the Human Diet.* NRC. National Academy Press, Washington, D.C., 1996.
4. Ames, B.N., Magaw, R., Gold, L.S. *Science* **1987**, *236*, 271-280.
5. Gold, L.S., Slone, T.H., Stern, B.R., Manley, N.B., Ames, B.N. *Science* **1992**, *258*, 261-265.
6. Winter, C.K., Francis, F.J. *Food Technol.* **1997**, *51*, 85-92.
7. Covello, V.T., Sandman, P.M., Slovic, P. *Risk Communication, Risk Statistics, and Risk Comparisons: A Manual for Plant Managers.*; Chemical Manufacturers Association, Washington, D.C., 1988.
8. Grisebach, H., Ebel, J. *Angew. Chem. Int. Ed. Engl.* **1978**, *17*, 635-647.
9. Beier, R.C., Nigg, H.N. In *Foodborne Disease Handbook: Diseases Caused by Hazardous Substances.* Hui, Y.H., Gorham, J.R., Murrell, K.D., Cliver, D.O., eds., Marcel Dekker, New York, N.Y., 1994; Vol. 3; pp. 1-186.
10. Beier, R.C., Oertli, E.H. *Phytochemistry* **1983**, *22*, 2595-2597.
11. Veech, J.A. *J. Nematol.* **1979**, *11*, 240-246.
12. Lord, K.M., Epton, H.A.S., Frost, R.R. *Plant Pathol.* **1988**, *37*, 385-389.
13. Wu, C.M., Koehler, P.E., Ayres, J.C. *Appl. Microbiol.* **1972**, *23*, 852-856.
14. Dercks, W., Trumble, J., Winter, C. *J. Chem. Ecol.* **1990**, *16*, 443-454.

15. Hsieh, D.P.H., Gruenwedel, S.H.O. In: *Chemicals in the Human Food Chain*, Winter, C.K., Seiber, J.N., Nuckton, Eds.; Van Nostrand Reinhold, New York, N.Y., 1990 pp. 239-267.
16. Bezuidenhout, S.C., Gelderblom, W.C.A., Gorst-Allman, C.P., Horak, R.M., Marasas, W.F.O., Spiteller, G., Vleggaar, R. *J. Chem. Soc. Chem. Commun.* **1988**, 743-745.
17. Rheeder, J.P., Marasas, W.F.O., Thiel, P.G., Sydenham, E.W., van Schalkwijk, D.J. *Phytopathology* **1992**, *82*, 353-357.
18. Caldas, E.D., Jones, A.D., Ward, B., Winter, C.K., Gilchrist, D.G. *J. Agric. Food Chem.* **1994**, *42*, 327-333.
19. Caldas, E.D., Jones, A.D., Winter, C.K., Ward, B., Gilchrist, D.G. *Anal. Chem.* **1995**, *67*, 196-207.
20. Dalcero, A., Combina, M., Etcheverry, M., Chulze, S., Rodriguez, M.I. *Food Additives and Contaminants* **1996**, *13*, 315-320.
21. Chourasia, H.K. *Nat. Acad. Sci. Letters (India)* **1992**, *15*, 243-246.
22. El-Kady, I.A., El-Maraghy, S.S.M., Abdel-Mallek, A.Y., Hasan, H.A.H. *Zentralbl. Mikrobiol.* **1993**, *148*, 549-557.
23. Arino, A.A., Bullerman, L.B. *J. Food Prot.* **1993**, *56*, 718-721.
24. Calori-Domingues, M.A., Fonseca, H. *Food Additives and Contaminants* **1995**, *12*, 347-350.
25. Omar, S.A., Hahmoud, A.L.E. *Mycoses* **1995**, *38*, 93-96.
26. Widstrom, N.W. *J. Environ. Qual.* **1979**, *8*, 5-11.
27. Gianessi, L.P. Crop Protection Issues Paper #1, National Center for Food and Agricultural Policy, Washington, D.C., 1997.
28. El-Morshedy, M.M.F., Aziz, N.H. *Bull. Environ. Contam. Toxicol.* **1995**, *54*, 514-518.
29. Gareis, M., Ceynowa, J. *Z. Lebensm. Unters. Forsch.* **1994**, *198*, 244-248.
30. Trumble, J.T., Millar, J.G., Ott, D.E., Carson, W.C. *J. Agric. Food Chem.* **1992**, *40*, 1501-1506.
31. Nigg, H.N., Strandberg, J.O., Beier, R.C., Petersen, H.D., Harrison, J.M. *J. Agric. Food Chem.* **1997**, *45*, 1430-1436.
32. Komives, T., Casida, J.E. *J. Agric. Food Chem.* **1983**, *31*, 751-755.

Chapter 13

Importance of Pesticides in Integrated Pest Management

D. Raymond Forney

DuPont Agricultural Products, Chesapeake Farms Project, 7321 Remington Drive, Chestertown, MD 21620

Prior to the widespread use of chemical pesticides, growers protected their crops and enhanced yields by using manual labor, cultural and biological methods, mechanical cultivation, and crop rotation. In the mid-1900s, efficient and economical pesticides were introduced, allowing growers to take a preventive approach in the battle against pests and disease. However, widespread use of persistent pesticides resulted in concerns about pest resistance and environmental contamination.

The practice of integrated pest management (IPM) involves the prescriptive use of pesticides along with many other pest management tools. A key to the success of IPM is the development of pesticides that are selective to specific pests, present favorable safety and environmental profiles, and are effective at low application rates. Combining new chemistries and application methods with other control methods provides growers with effective tools for pest management while minimizing risk to humans and the environment.

The Food and Agriculture Organization's (FAO) International Code of Conduct on the Distribution and Use of Pesticides states that

> "Integrated Pest Management (IPM) means a pest management system that, in the context of the associated environment and the population dynamics of the pest species, utilises all suitable techniques and methods in as compatible a manner as possible and maintains the pest populations at levels below those causing economically unacceptable damage or loss" (*1*).

In simpler terms, the grower using IPM combines cultural, biological, strategic, and chemical tools to obtain the most cost-effective control of diseases, insects, and weeds at the least risk to humans and the environment.

IPM is not a new concept. For centuries, growers used all the tools they had available to control the pests that destroyed their crops and threatened their families. But with the advent of the Industrial Revolution and the subsequent shift from subsistence farming to commercial farming, the need for more effective pest control became critical. The destruction of crops by a pest infestation meant more than hardship for a farm family; it could mean widespread famine. Such agricultural disasters provided the impetus that led to the development of synthetic pesticides.

Inexpensive, easy to apply, and extremely effective, early synthetic pesticides significantly reduced insect-borne disease and increased agricultural productivity (*2*). By the 1950s and 60s, the agriculture industry had come to regard these materials as the single solution to pest control problems. Pesticide use became widespread and sometimes indiscriminate. In addition, the understanding of the science of pesticides had not advanced as far as the ability to produce them. Ensuing problems with pest resistance and human health and environmental effects served as a wake-up call not only for the pesticide industry, but for the grower and consumer as well.

Today, pesticides are no longer seen as the single solution, but as an important part of the overall solution. Although synthetic pesticides remain the primary means of controlling pests (*3*), modern IPM systems take advantage of the benefits of various pest management practices—cultural, strategic, biological, and chemical—while minimizing the risks. Significant new research is resulting in pesticides that are selective, effective at low rates, and present low risk to humans and the environment. New application techniques mean that pesticides can be placed exactly where they are needed and when. These new technologies and our increased knowledge of the often subtle relationships between humans and the environment will enable growers to use IPM to tailor pest management to each individual situation.

This paper discusses the historical use of pesticides and how it led to current products, the role of these products in IPM systems, and future technologies that will help define agriculture in the coming century.

Need for Pest Management

The world population is expected to reach six billion by year 2000 and exceed 10 billion in the 21st century (*4*). Until recently, food production has been able to keep up with the demand, mostly because of the "Green Revolution." Crop rotation, use of fertilizers and chemical pesticides, expanded irrigation, and the development of disease-resistant crops have all contributed to the dramatic increase in agricultural yields during the 20th century. However, this picture may be changing. Most of the population growth is taking place in developing countries where, at the same time, the arable land per capita is dwindling (Figure 1). In 1961, developing countries averaged about 0.3 hectare of arable land per person. By 1992, the average had declined to less than 0.2 hectare per person (*5*).

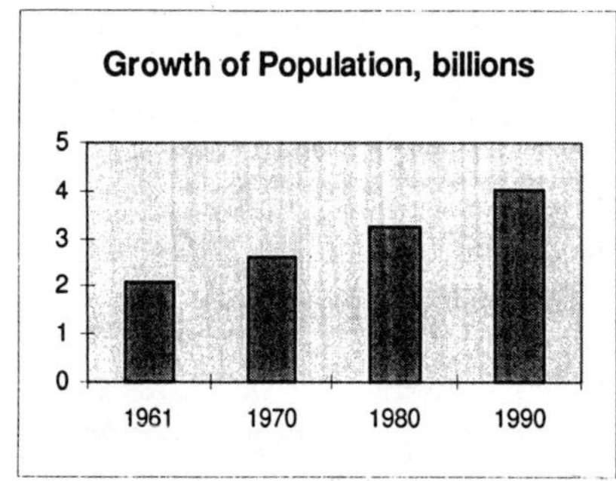

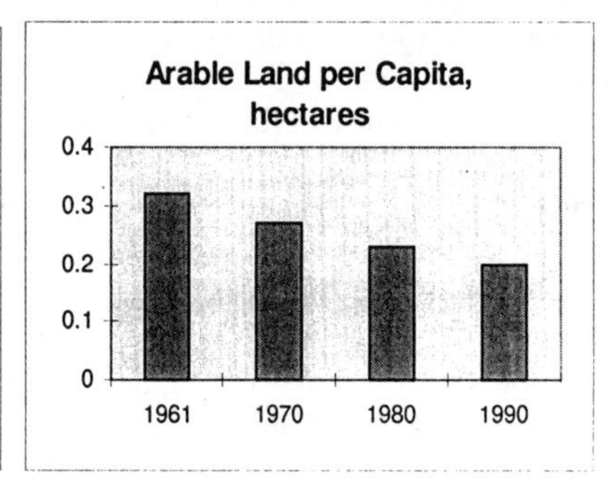

[a]Adapted from ref. 5.

Figure 1. Population and Arable Land in Developing Countries

Other problems are contributing to the inability of developing countries to support their populations. Destructive farming and fishing practices limit the productivity of land and thereby limit the sustainability of the agriculture that exists. According to the FAO, sustainable agriculture is defined as agriculture that conserves land, water, and plant and animal genetic resources, does not degrade the environment, and is economically viable and socially acceptable (*5*). In addition, the arable land that exists in developing countries is becoming concentrated in fewer farms. Although such land concentration can increase yields, it often results in export crops being grown at the expense of food crops for domestic use. Currently, about two billion people around the world do not have access to safe and nutritious food, either because they cannot grow enough food themselves, or they cannot afford to buy it (*5*).

In order for growers to feed more people on less available land, agriculture needs to be as efficient as possible, producing high-quality abundant food at low cost. The more efficient agriculture is in a country, the higher its standard of living will be. Table I shows the effects of agricultural efficiency on the affluence of several countries. Ironically, as yields increase, the need for pest management increases also. This is because increased crop density, shortened period between crops, monoculture, and increased use of fertilizers encourage pests and diseases (*6*). The development of new technologies in all aspects of pest management will be crucial not only to keeping agricultural yields high enough to feed a growing population, but also to making those yields sustainable in the future.

Table I. Role of Agriculture in National Affluence

Country	% of Workforce in Agriculture[a]	% of Income Spent on Food[b]	Life Expectancy[b]	Infant Mortality per 1,000[b]
U.S.A.	2.2	6.8	76	9
France	5.0	11.3	77	7
Russia	13.5	13.2	69	20
Brazil	23.7	24.5	66	58
India	66.2	35.3	60	90

[a]World Bank, 1994 data.

[b]Food and Agriculture Organization of the United Nations, 1994 data.

Early Pest Control

Growers have been trying to control the pests that diminish crops and carry disease for centuries. As early as 2500 BC, the Sumerians were using sulfur compounds to control insects and mites. By 1200 BC, the Chinese were using plant-derived insecticides for seed treatment and fumigation. Almost 800 years later, the Chinese had developed their knowledge of insects and plants sufficiently to establish predatory ants in citrus orchards to control caterpillars and large boring beetles (*7*). They used chalk and wood ash for prevention and control of indoor and stored-product pests, and mercury and arsenic compounds to control body lice. The Chinese also recognized the value of adjusting crop planting times to avoid pest outbreaks (*8*).

Early methods of weed control were manual. Until about 3000 BC, weeds were controlled by hand-pulling. Wooden implements were used from 3000 to 2000 BC and then were supplemented by hand sickles and the first wooden plow about 1000 BC. A wooden spiked-tooth harrow had been invented by 500 BC (*7*).

With the discovery of the compound microscope in the 17th century, pest control had the advantage of an improved understanding of the biology of insects and plants. Various botanical insecticides, such as pyrethrum, quassia, and tobacco leaf infusion, became available in Europe. Several of these insecticides are still in use today. At the same time, the dangers of early inorganic pesticides were recognized when France banned the use of arsenic and mercury steeps for seed treatment in 1786 (*8*).

A major shift in agriculture coincided with the Industrial Revolution in Europe in the mid-1700s. Farming was no longer a subsistence operation feeding only the farm family, it was a commercial venture, feeding the thousands moving to urban locations and taking on industrial work. Agricultural yields rose as land was redistributed and planting acreages were expanded. New agricultural practices such as manuring and crop rotation also boosted yields. Growers began cropping in rows to enable weed control by horse-drawn hoe (*8*).

As these large-scale, commercial farming operations continued to expand, the need for effective, efficient pest control became crucial. Several major agricultural disasters in the mid-1800s were attributable in part to this expansion and to the introduction of new pests from other locations. The potato blight in Ireland, England, and Belgium; the epidemic of the fungus leaf spot disease of coffee; and the infestation of European vineyards by the American pest, the grape phylloxera, caused scientists to look seriously at the systematic development of pest control practices and materials. Breakthroughs included the use of host plant resistance and grafting against the grape phylloxera and the development of Bordeaux mixture (hydrated lime plus copper sulfate) and Paris Green (copper acetoarsenite) as fungicides. At about the same time, biological controls were being studied. The vedelia beetle (*Rodolia cardinalis*) and a parasitic fly (*Cryptochaetum iceryae*) were imported from Australia for control of cottony cushion scale in California. The beetle proved to be successful in practically eradicating cottony cushion scale. The parasitic fly was used effectively for pest control in southern California (*8*).

By the early 20th century, it was discovered that many serious diseases could be controlled by controlling the insects that carried them, such as rat fleas, a vector or carrier for the plague, and mosquitoes, a vector for malaria and yellow fever. In addition to existing chemical and biological controls, entomology and other textbooks published during the early 1900s promoted crop rotation, arrangement of planting times, fertilization, and proper soil preparation as means of pest control.

Advent of Synthetic Pesticides

In the 1920s and 30s, pest control still relied heavily on a combination of tactics to control pests. This situation changed dramatically with the outbreak of World War II. Tropical warfare and the accompanying insect-carried diseases, such as typhus, encephalitis, dengue, and malaria, wreaked havoc on all combatants. The United States conducted an intense effort to screen hundreds of compounds from around the world for insecticidal activity. One of these compounds, dichloro-diphenyl-trichloroethane (DDT) has been credited with limiting the casualties of Allied forces by controlling disease-carrying insects and practically eradicating pests such as malaria-carrying mosquitoes from countries where such pests presented a serious health threat (*7*).

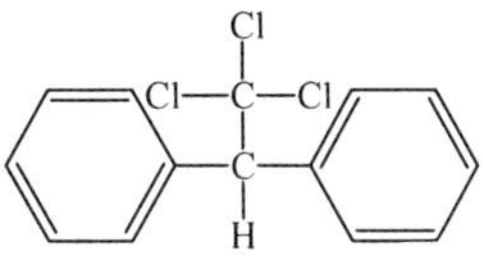

DDT

Later discoveries included organophosphates from Germany and carbamates from Switzerland, both originally developed for control of disease-carrying insects, later to be used extensively in agriculture. These materials were inexpensive, effective, easy to apply, and widely toxic. Although the risks of widespread use of some of the early synthetic pesticides would be discovered later, the benefits in terms of lives saved, reduced suffering, and improved agriculture should not be underestimated.

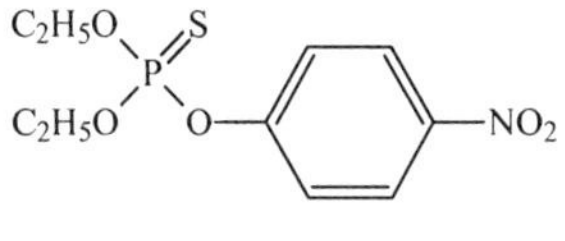

Parathion

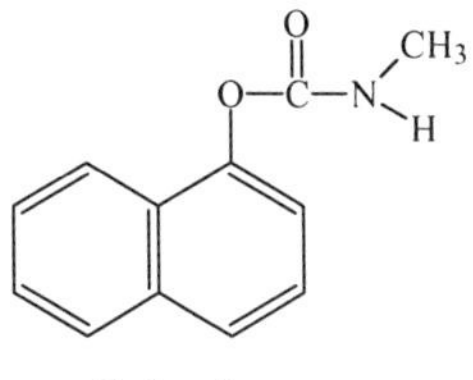

Carbaryl

Some specific pesticide developments are discussed below. Table II lists a chronology of pesticide chemistries beginning in the 1960s.

Herbicides. The advent of synthetic herbicides was largely responsible for today's machine-harvested crops (*9*). In 1994, there were more than 125 different active ingredients and 100 premixes of these ingredients marketed in the U.S. for use as herbicides (*10*). Many herbicides are selective, controlling one group of plants while not affecting nontarget plants. Products such as 2,4-D, fosamine, dicamba, and picloram control many broadleaf weeds, but do not damage grasses, sedges, or ferns. Other herbicides, such as quizalofop, control grasses but do not affect broadleaf plants. This selectivity can be modified by changing the application rate

Table II. Chronology of Pesticide Introductions

Decade	Insecticides	Herbicides	Fungicides
1960s	Organophosphates: fonofos, trichloronate	Amides: propachlor, alachlor, butachlor	Oxathlines: carboxin
	Carbamates: carbofuran, aldicarb, methomyl	Ureas: methabenzthiazuron, chlortorolon	Guanidines: triforine, guazatin
	Pyrethroids: resmethrin	Toluidines: trifluralin	Organophosphates: edifenphos
	Formamidines: chlordimeform	Diazines: bentazone, oxadiazon	Pyrimidines: ethirimol
			Morpholines: tridemorph
			Benzimidazoles/thiophanates: thiophanate-methyl
1970s	Pyrethroids: permethrin, cypermethrin, deltamethrin, fenvalerate	Triazines: cyanazine, metribuzin, metamitron	Organophosphates: pyrazophos, fosetyl-al
	Organophosphates: terbufos, methamidophos, acephate	Carbamates: emtam+safener, phenmedipham	Pyrimidines: fenarimol
	Carbamates: bendiocarb, thiofanox	Ureas: isoproturon	Benzimidazoles/thiophanates: benomyl, carbendazim
	IGRs: methoprene, diflubenzuron	Toluidines: pendimethalin	Dicarboximides: procymidone, iprodione, vinclozolin
	AChE receptor blockers: cartap	Diazines: methazole	Imidazoles/triazoles: triadimefon, propiconazole
		Diphenyl ethers: dichlofop, oxyfluorfen	Phenylamides: metalaxyl
		Pyridine derivatives: clopyralid	Others: tricyclazole
		Cyclohexanediones: alloxydim	
1980s	Pyrethroids: flucythrinate	Carbamates: prosulfocarb	Morpholines: fenpropidin
	Procarbamates: carbosulfan, thiodicarb	Diphenyl ethers: acifluorfen, aclonifen	Imidazoles/triazoles: prochloraz, bitertanole, flusilazole, tebuconazole
	IGRs: phenoxycarb	Pyridine derivatives: fluroxypyr	Phenylamides: oxadixyl
	Microbials: BT, BTI, *Bacillus sphaericus*	Cyclohexanediones: sethoxydim, tralkoxydim	Others: probenazole, pyrifenox, fenpiclonil
	AChE receptor blockers: bensultap	Aryloxyphenoxypropionic acid: fluazifop, fenoxaprop	
	GABA agonists: milbemycin, avermectin	Sulfonylureas: chlorsulfuron, bensulfuron methyl, tribenuron methyl, thifensulfuron	
	Miscellaneous: AMDRO, cyromazine	Amino acid derivatives: glyphosate, glufosinate	
		Imidazolinones: imazaquin, imazamethabenz, imazethapyr	

[a]Data from ref. 9 and ref. 16.

of the product or the season of application. On the other hand, broad-spectrum herbicides, such as glyphosate, will control most vegetation. Sulfonylurea herbicides are recent introductions that are effective at very low rates (*9*). In this respect, sulfonylurea herbicides represent a new generation in herbicide development. In general, the active ingredients in herbicides are of relatively low toxicity to humans and animals (*10*).

Insecticides. Chlorinated hydrocarbons such as DDT represent the first generation of insecticides. Although widely used in the mid-1900s, many of these materials are either no longer used or are restricted because of problems that developed with resistance, persistence, and toxicity. Organophosphates, the next generation, are widely used and among the most important crop protection products. More than 75 of these compounds are in use as fast-acting systemic insecticides and soil insecticides (*9*). Compared to chlorinated hydrocarbons, organophosphates are less environmentally persistent, more biodegradable, less subject to biomagnification, and usually unstable in the presence of sunlight. Although development of pest resistance occurs, the problem is less serious than that with chlorinated hydrocarbons (*11*). A limiting factor on the use of organophosphates is their adverse effects on beneficial insects.

Carbamates are mainly used as soil insecticides and for controlling pests resistant to chlorinated hydrocarbons and organophosphates. They have many of the same benefits and risks as organophosphates. Benzoyl ureas are insect growth regulators that inhibit the development of insect larvae. They are slow-acting and are usually combined with other insecticides (*9*).

There are other groups of insecticides, including several biologically derived materials. One of the most recent introductions is the spinosyns, which are metabolites of soil microorganisms (*12*). The spinosyns show activity against selected insect groups, such as lepidoptera and diptera, at low application rates and low toxicity to humans and the environment.

Fungicides. Surface-protective fungicides, which include inorganic fungicides and organics such as dithiocarbamates and dicarboximides, cannot penetrate plant tissue and eliminate existing fungal infections. They are applied in advance of infection and serve to protect the plant. Inorganic fungicides include sulfur and copper compounds that have been used for several centuries. Dithiocarbamates were developed in the 1930s and are still widely used in cereals and specialty crops. The dicarboximides, such as captan and folpet, show an unusual stability to fungal resistance, probably due to their multiple modes of action (*13*).

Systemic fungicides can penetrate plant tissue and eradicate or control existing infection, but resistance development is more of a problem than with surface protectants. Organophosphates show specific action against *Pyricularia oryzae* infection of rice. The pyrimidines are used for dressing seed, for soil treatment, and for foliar application for controlling powdery mildew. The benzimidazoles, including benomyl, are broad-spectrum materials with a specific mode of action. Phenylamides, such as metalaxyl, were introduced in 1977 and are used at low rates for selective control of *Oomycetes* infections. Although resistance can develop to compounds in this group, it can be avoided by using combinations of fungicides and suitable treatment sequences (*9*, *13*).

Reliance on Synthetic Pesticides and Resulting Problems

The high yields of modern agriculture and the resulting high standard of living have been directly facilitated by innovations in agricultural technology, including chemical pesticides, high-yield variety crops, fertilizers, and mechanical and energy inputs (*14*). However, the benefits of widespread adoption of these technology-driven innovations have been accompanied by some undesirable consequences during the evolution of modern agriculture (*15*). It is these consequences that have spurred major recent advances in pesticide development and adoption of the concept of integrated pest management.

Development of Resistance. As exposure to a pesticide increases, individual pests become resistant to the material and pass this resistance on to the next generation. Gradually, an increasing percent of the pest population becomes resistant to control products that originally worked very well. Insects in particular have a remarkable ability to adapt and evolve. Resistance to insecticides was first recognized almost 76 years ago, although the major incidences of resistance have occurred with the advent of synthetic insecticides in the last 40 years (*16*). One of the most notable incidences is the development of resistance in pests of cotton. Chlorinated hydrocarbons had been found to control the boll weevil, a pest that had historically severely limited cotton production. However, widespread use of these products led to resistance development. Later, low rates of organophosphate would provide good control of the boll weevil, but these rates were insufficient to control the *Heliothis* complex, which was quickly becoming a major threat to cotton. As rates were increased to achieve *Heliothis* control, resistance to organophosphates became a problem.

Although resistance development has been most obvious with insecticides, other pesticides are affected as well. Resistance in plant pathogens was first detected 44 years ago. More recently, weeds are exhibiting resistance to herbicides (*16*).

The response to pesticide resistance has traditionally been to apply the pesticide at a higher rate, increase the frequency of application, or change pesticides. The first two solutions are self-defeating, as they tend to exacerbate the problem of resistance. The third solution may ultimately cause its own resistance problems unless there is a basic change in mode of action and treatment strategy (*17*).

Recent scientific advances have made considerable progress toward combating resistance development. The likelihood of resistance is one of the most important considerations during the development of a new pesticide. For existing pesticides, there are several tactics that can be used to suppress resistance:

- pesticide mixtures using different modes of action
- pesticide rotation using different modes of action

- implementing other IPM strategies
- controlled application and application timing

Ultimately, success will come from a better understanding of the biochemistry and genetics of resistance development.

Target Pest Resurgence and Secondary Pest Outbreak. Target pest resurgence occurs when the target pest, whose numbers were decreased by application of a pesticide, suddenly experiences a dramatic increase. This is usually due to the detrimental effects of a pesticide on the natural enemies of the pest. As the target pest population goes down, the natural enemies lose their food source and either die, move on to other areas, or fail to reproduce. In addition, the pesticide may be killing the natural enemy as well as the target pest. Because the initial control of the target pest was probably due to a combination of the activities of the pesticide and the natural enemy, loss of the natural enemy for any reason leaves an environment that is more favorable for the pest to flourish.

Secondary pest outbreak occurs when pests that had not been targeted by the pesticide application suddenly become a problem when the natural enemy population is diminished upon treatment of the target pest. Although these secondary pests always had the potential to be primary pests, their status changed when their environment was disrupted by loss of natural enemies or even loss of the primary target pest, with which the secondary pest had competed in the past.

A secondary pest outbreak occurred in California when DDT was applied in citrus orchards for control of pests. The imported vedalia beetle, which had kept the cottony cushion scale under control for many years, proved to be very susceptible to DDT. Once the population of the vedalia beetle diminished, the scale population soared. It was only after DDT applications were adjusted and the vedalia beetle was reestablished that the cottony cushion scale was again brought under control (*8*).

Health and Environmental Effects. Pesticides are biologically active materials and as such pose a potential risk to humans and the environment as well as to the pests they target. Adverse effects caused by pesticides to the environment and to human health are well documented (*18*, *19*), although not wholly agreed upon (*20*). That such risks need to be a significant consideration during the development of a pesticide is reflected in the recent introductions of low-rate, selective, nonpersistent pesticides. These new products are in direct response to lessons learned from the widespread use of some of the early synthetic pesticides.

Again, DDT is an example. Widely used, broadly toxic, and persistent in the environment, DDT showed far-reaching effects on the environment. Areas where the pesticide was never applied showed evidence of DDT residues. DDT was also found to biomagnify as it progressed through the food chain, such that a level of $2x10^{-6}$ ppm in a body of water could increase to 10 ppm in fish (*2*). Other early pesticides had similar effects.

Acute human health effects of pesticides are difficult to assess because of the lack of reliable statistics. The most serious problems are in developing countries where there is evidence that a major portion of acute poisonings are due to suicides. Approximately one-fourth of casualties are due to accidents in the home or workplace (*21*). The problem is less serious in developed countries where applicators are better educated and the use of pesticides that are considered to have a high acute risk is restricted to trained professionals that must be certified. However, both in developed and developing countries, most accidental poisonings are due to misuse of the product.

Chronic effects of pesticide residues on human health is currently the subject of intense debate. While advocates on one side of the issue maintain that there are dangerous levels of pesticide residues on food (*22*), advocates on the other side question the science of some of the residue studies, particularly when extremely large amounts of food would need to be consumed before humans would actually be exposed to the residue level studied. At the least, a balanced approach to the issue is being sought (*23*).

To address the concerns of risk to the environment and to human health, pesticides are required to undergo stringent evaluation before they can be used. It typically takes 8 to 10 years and $50 million in research and development testing before a new pesticide has completed the tests necessary for submission for registration by the Environmental Protection Agency.

The Development of IPM

As mentioned earlier, integrated pest management is not a new idea. However, it is only within about the last 40 years that it has developed from a rather casual combination of control measures to a cohesive pest management strategy. And it is only in the last 20 years that it has been extensively researched and federally funded.

Contrary to the approach commonly taken in the mid-1900s, when widespread pesticide use was advocated, IPM seeks to suppress pest populations to avoid economic losses rather than eradicate the pest entirely (*24*). As the USDA definition of IPM states, accomplishing this goal includes the prescribed use of pesticides:

> "IPM is a management approach that encourages natural control of pest populations by anticipating pest problems and preventing pests from reaching economically damaging levels. All appropriate techniques are used such as enhancing natural enemies, planting pest-resistant crops, adapting cultural management, and using pesticides judiciously" (*25*).

Crucial to the success of IPM and to the decision as to what control methods will be used is the determination of when and if pest control is necessary. The economic threshold is the level of pest infestation at which physical crop damage and revenue losses become excessive and the benefits of pest control exceed the cost and potential risks. In other words, what level of pest infestation can be tolerated before losses for the grower and the consumer become too high?

Scouting (sampling for pests and beneficial organisms in the field) is the primary way that growers monitor pest populations to determine if an economic threshold is being approached. Scouting also helps determine the appropriate method or combination of methods of pest control depending on the severity of infestation and the type of pest. Computer modeling using weather conditions and other factors can also help predict the severity and timing of a pest infestation (*26*) and the best methods of control.

If all of the possible IPM techniques are reviewed, the picture is a complex one. Figure 2 shows the various tools available (*27*). However, this picture can be simplified by placing the tools into specific categories:

Biological Controls. Biological controls include natural enemies (parasites, predators, and insect pathogens), semiochemicals (pheromones and feeding attractants), and biopesticides (*26*). Biological controls present, in general, lower risk to humans and the environment than conventional pesticides. However, biological pesticides are not entirely without risk. There is relatively little documentation in the scientific literature regarding studies of either health or environmental effects of biological pesticides compared to the studies that have been done on synthetic pesticides. However, several studies have shown evidence of allergic reactions, adverse effects on beneficial organisms, and resistance development (*28*).

While natural enemies and biopesticides are used to kill pests, semiochemicals such as pheromones are usually used to monitor for pests or to attract them to baited areas that contain pesticides (*29*). Biological controls provide, in general, a partial solution to pest problem and are usually integrated with other control methods (*30*). Although natural enemies can provide long-term control of key pests, commercial biocontrol products generally have a narrow target range, act slowly, have a short field persistence and shelf life, and their window of application is relatively narrow. Pest control using a combination of natural enemies and biopesticides with synthetic pesticides is showing increasing promise because of the improved selectivity of synthetic pesticides (*31*).

Cultural Controls. Cultural controls include cultivation, sanitation, and crop rotation (*26*), among others. These techniques alter the pest's environment, making it less conducive to reproduction and feeding and sometimes more favorable for the pest's natural enemies. Cultivation kills pests by injury, starvation, desiccation, and exposure. It does have serious side effects in the loss of organic matter and wind and water erosion of soil. Sanitation involves removing and destroying diseased plants, insect breeding areas, overwintering sites, and food sources such as garbage and other wastes. It is particularly important in industrial and home pest control. Crop rotation controls pests by following the crop of one family with a crop that is not a host for the pest concerned (*2*).

Strategic Controls. Some strategic controls include planting location, planting date, and timing of harvest. The purpose of strategic controls is to give the crop a competitive advantage over pests. Planting either before or after emergence of insects can prevent serious damage. Growers can time harvests to avoid predicted infestations of pests. They can also avoid planting locations that are prone to pest

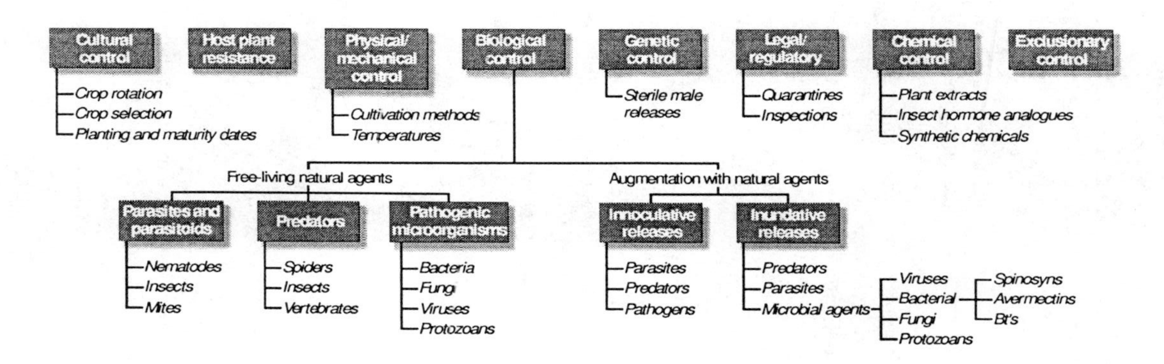

[a]Reproduced with permission from ref. 27. Copyright 1997 DowElanco.

Figure 2. Tools Available for Integrated Pest Management[a]

infestation and disease (*8*). Host plant resistance is sometimes considered a strategic control, sometimes a cultural control. Many plants have a natural resistance to insects and disease. Scientists and growers have taken advantage of this characteristic by hybridizing and selecting plants known to have host resistance (*32*). Control of the wheat stem sawfly, spotted alfalfa aphid, and European corn borer are notable success stories (*33*).

M. Kogan offers a more comprehensive review of pesticide alternatives (*34*). Although a complete review of biological, cultural, and strategic controls is outside the purview of this paper, it is clear that the incomplete applicability or performance of these techniques in modern agriculture creates opportunity for other technologies.

Chemical Controls. Despite the availability of alternative methods of pest control, use of chemicals is still an important part of an IPM system for several reasons. Other methods of pest control are not available to all growers, they may be very expensive to use, and they may not be effective enough. In fact, it is unlikely that present levels of production could be sustained without the prescribed use of chemical pesticides. However, it is vital that such pesticides have as many of the following characteristics as possible: efficacy at low rates; selective toxicity to specific pest species; low toxicity to beneficial organisms, humans, and the environment; good biodegradability; and low mobility to groundwater. Moreover, they should be used where and when they are actually needed, and label directions should be followed at all times. Later in this paper, current trends in pesticide development that reflect these requirements will be discussed.

The Current Status of IPM in the U.S.

In 1993, the Clinton Administration pledged that 75% of croplands would be managed with IPM systems by the year 2000 (*35*). One of the challenges to accomplishing this is acceptance by the grower. If IPM is found to or is perceived to increase costs, growers will be reluctant to attempt it. Growers also cite other concerns as obstacles to the adoption of IPM and the use of economic thresholds, including appearance of fields, weeds interfering with harvest, weed seed production, and the time required to scout fields (*36*). Because growers tend to associate pesticide use with reduction of yield variability and therefore income variability, they often avoid integrated systems, which can show a great deal of variability in returns. This is despite the fact that the ultimate returns of an integrated system may be greater (*37*). Educating growers as to the advantages and economic benefits of IPM will be essential to its success.

Determining the economic advantages of IPM is a complex problem, but it is being addressed. A 1994 literature review assessed the current state of knowledge on the economic evaluation of IPM programs (*38*). The review also compiled the results of farm-level evaluations of eight commodity crops: cotton, soybeans, vegetables, fruits, peanuts, tobacco, corn, and alfalfa. In general, the authors found that IPM programs reduced pesticide use, production costs, and risks and resulted in high net returns to growers. Scientists are also trying to evaluate the environmental benefits of IPM from an economic standpoint (*39*).

A study of the current status of IPM programs in the U.S. was sponsored by the Economic Research Service of the USDA (26). The study summarizes the use of IPM programs by growers as of 1991 and will provide a baseline for measurement of the success of the "75% IPM by 2000" goal. Figures 3 through 5 give the evaluation of IPM use in fruits and nuts, vegetables, and corn in 1991, respectively. The study concluded that although IPM acceptance has improved over the last 20 years and now reflects about half of fruit and nut, vegetable, and field crops, levels of adoption still vary widely, mostly due to pest severity, lack of effective and economical alternatives to conventional pesticides, inadequate knowledge of IPM, and the higher managerial input required.

The Changing Role of Pesticides in IPM

As mentioned earlier, pesticides are now considered a part of the solution to pest control, rather than the entire solution. But they are a vital part. A study undertaken by Texas A&M University in cooperation with the Tennessee Valley Authority sought to measure the impact of eliminating synthetic pesticide use (*40*). Table III summarizes the findings of this study for several crops. The results show that the elimination of pesticides would not allow us to sustain agriculture at its current levels, let alone increase productivity.

Table III. The Effect of Lack of Pesticides on Crops[a]

Crop	Loss w/o Pesticides[b]
Corn	32%
Soybeans	37%
Wheat	24%
Barley	29%
Cotton	39%
Rice	57%
Peanuts	78%

[a]Data from ref. 40.
[b]Herbicides, insecticides, and fungicides

Rather than eliminating pesticides, the answer to concerns about detrimental effects of pesticides on human health and the environment is to increase the benefits of pesticides and lower the risks, while using pesticides to augment the alternative methods of pest control offered by IPM. One of the major advances in pesticide development is increased efficacy at lower rates. Table IV gives some examples of how application rates of pesticides have decreased over the last 30 years. When pesticides are used in smaller quantities there is less likelihood of environmental contamination. The environmental situation is improved further if they are also biodegradable and nonpersistent.

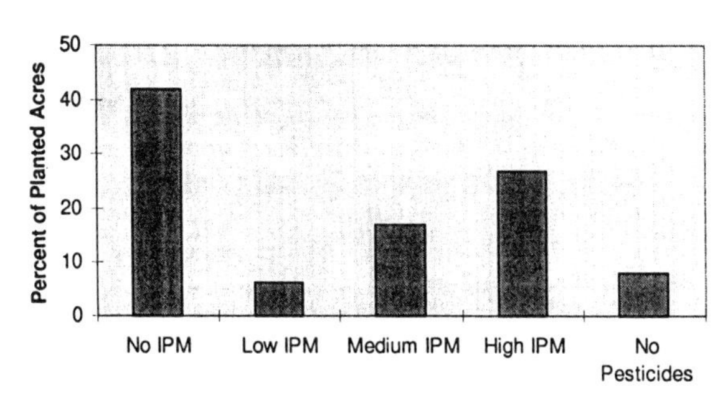

[a]Adapted from ref. 26, p. 10.

Figure 3. Adoption of IPM in Fruits and Nuts, 1992[a]

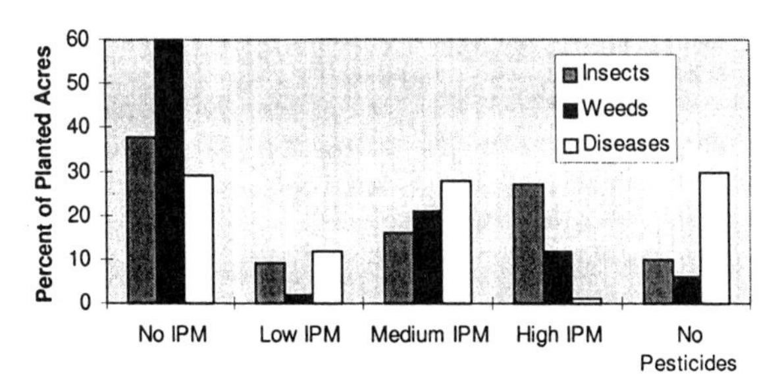

[a]Adapted from ref. 26, p. 20.

Figure 4. Adoption of IPM in Vegetables, 1992[a]

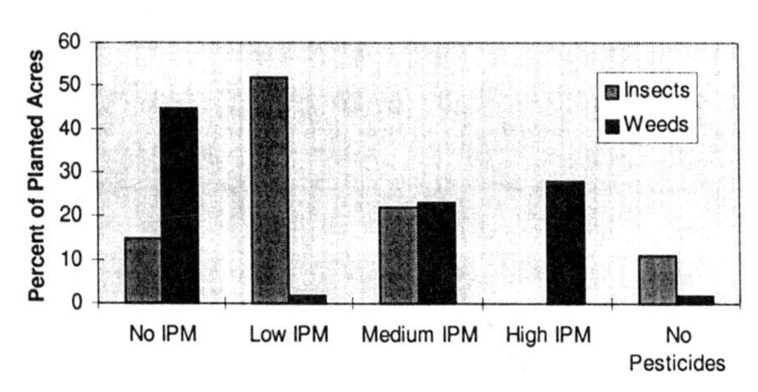

[a]Adapted from ref. 26, p. 23.

Figure 5. Adoption of IPM in Corn, 1992[a]

Table IV. Application Rates of Selected Herbicides, Insecticides, and Fungicides

Active Ingredient	Group	Year of Introduction	Typical Application Rate (g/ha)
Herbicides			
Alachlor	Amide	1966	3,000
Chlorsulfuron	Sulfonylurea	1982	20
Imazethapyr	Imidazolinone	1987	100
Insecticides			
Aldicarb	Carbamate	1965	2,500
Diflubenzuron	Benzoyl urea	1977	100
Deltamethrin	Pyrethroid	1982	20
Fungicides			
Mancozeb	Dithiocarbamate	1961	1,200
Carbendazim	Benzimidazole	1973	250
Difenconazole	Triazole	1991	100

[a]Adapted from ref. 9, p. 55.

Many pesticides being developed today are designed to have a mode of action that is very selective to the pest to be controlled, lessening potential side effects to humans and beneficial organisms. Understanding the mode of action of pesticides is also important in managing resistance development. The development of new active ingredients with differing modes of action or switching from one mode to another reduces the risk that resistance will develop.

All of the above considerations play a major role in the development of new pesticides and in the use of current materials. Evaluations of the safety and environmental effects of new compounds are made very early in the development process, and indications of unacceptable risk bring development to a halt. The extremely close scrutiny that the development and use of pesticides receive from the U.S. government is further insurance that new pesticides reflect greater selectivity and safety and that current pesticides are used properly.

Summarized below are a few examples of how new pesticides are offering increased benefits at lower risks and how these materials work hand-in-hand with the goals of IPM.

New Developments in Pesticides. Sulfonylureas have been under intense development for the last 20 years and are used extensively in cereal crops for control of grasses and broadleaf weeds. These materials provide the yield increases typically expected of chemical pesticides while reducing actual pesticide application by 95 to 99% compared to higher use rate alternatives (*41*). In a study of pesticide usage in Oklahoma wheat, sulfonylureas were credited with the decrease in pounds of active ingredient applied to Oklahoma wheat between 1981 and 1995

(603,150 to 263,400) and the increase in acreage treated (877,000 to 4,825,000) (*42*).

The safety characteristics of sulfonylureas are very favorable to applicators, field workers, and consumers. They show very low toxicity in a broad battery of acute and chronic dietary, dermal, inhalation, and eye tests in animals. The low application rates and the fast metabolism in crops leave virtually no residues in harvested grain or straw. In addition, standard acute and chronic toxicity testing has shown that sulfonylureas pose negligible toxicity risks to nontarget organisms. Because sulfonylureas have low volatility and good biodegradability, the chance of groundwater contamination and other environmental contamination is very low (*41*). Resistance development with sulfonylureas is similar to that of other herbicide classes and can be controlled in a similar manner.

Two other characteristics of sulfonylureas are important in relation to IPM systems: applicability to postemergence application and compatibility with minimum or no-till agriculture. Postemergence application (applying after emergence of weeds) provides the greatest flexibility in allowing economic thresholds to determine the need for pesticide use and avoiding preventive applications. Compatibility with minimum or no-till agriculture means that sulfonylureas will be effective in tillage regimes that minimize the chance of soil and organic matter loss through erosion (*41*).

Another example of the importance of pesticides with postemergence activity is the control of johnsongrass, a serious pest weed in soybeans. In the 1970s, the only control for johnsongrass was a double rate (1.5 to 2 kg/ha) of a dinitroaniline herbicide (trifluralin) applied to the soil with extensive tillage. This treatment was only marginally effective, required 2 years of continuous soybeans, precluded conservation tillage, and was a broadcast, insurance-type treatment. With the discovery of postemergence grass herbicides, such as quizalofop, growers could grow no-till soybeans, scout for johnsongrass, only treat those areas of their fields where the grass appeared, and use far less active ingredient (0.07 kg/ha).

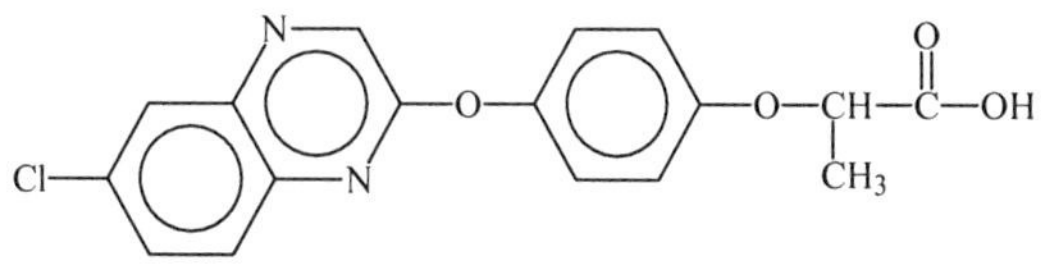

Quizalofop-P acid

A similar example involves the control of woolly cupgrass in corn. In the 1980s, a standard treatment used a soil application of EPTC at 3 to 4 kg/ha, again in a preventative mode with extensive soil tillage. With the introduction of nicosulfuron in the 1990s, growers can scout for emerging woolly cupgrass and control it where needed with a postemergence application of 0.035 kg/ha nicosulfuron.

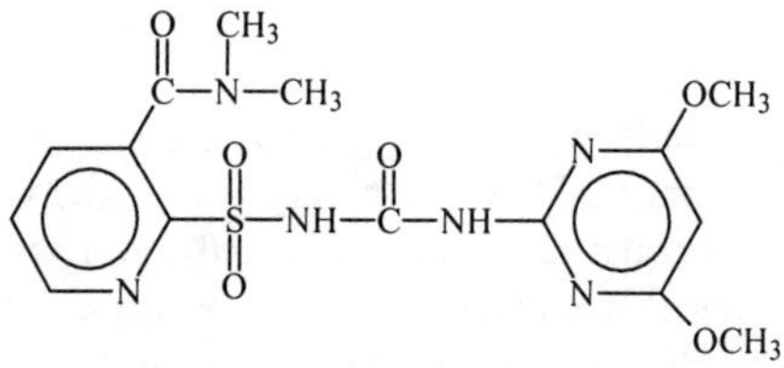

Nicosulfuron

Spinosyns, a new class of insecticides, are produced by fermentation of an actinomycete bacterium, *Saccharopolyspora spinosa*. This new class of insecticides is representative of the research being conducted today on synthetic pesticides based on biological materials. Spinosad, one of the metabolites in this class, is highly effective against lepidopterous pests at low use rates. It is not detrimental to typical beneficial predators and parasites, and it has been found to be nonpersistent and immobile in the environment. Spinosad also shows very low toxicity to humans and other nontarget organisms.

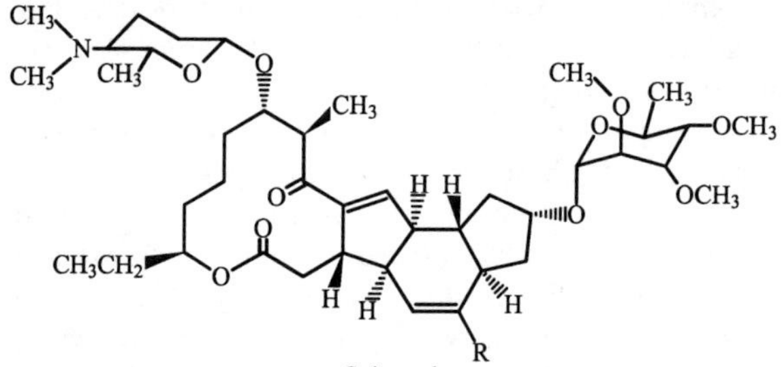

Spinosad

The use of spinosad in cotton is a good example of the contributions that a pesticide can make in an IPM program. Spinosad provides effective control of tobacco budworm, which has become increasingly resistant to pyrethroids. Because beneficial insects can tolerate applications of spinosad, they can extend the effective interval of control, offering the potential of fewer product applications compared to most standard treatments. Studies also suggest that when spinosad is used in the presence of beneficial insects, there will not be a significant flare-up of secondary pests such as beet armyworm, silver leaf whitefly, or soybean looper (*43*, *44*, *45*).

Other Trends in Pesticide Use in IPM Programs

Besides new chemistries, there are other developments in pesticide use and application that support and promote the goals of IPM:

Remote Sensing. Remote sensing is a tool by which aircraft or satellite-based devices capture spectral images of land, enabling assessments of crop productivity, crop and soil condition, and weed infestations. These systems have the potential of greatly increasing scouting efficiency for weeds while delivering that information to growers in a timely and usable manner. Although many of these systems are still under development, they are showing significant advantages in IPM programs. In one cited example, image-derived weed maps were combined with a geographic information system (GIS) model to determine the optimum herbicide mix and application rates for no-till corn. The treatment regime resulted in reductions of herbicide use by more than 40% (*46*). Without the availability of selective, postemergence herbicides, such an approach would not be possible.

Precision Applications. New application equipment is available that reduces the waste and potential for environmental contamination. One such system (SmartBox) delivers precise application rates regardless of speed. Conventional application equipment tends to deliver more product as the equipment speed decreases, at times resulting in more product being applied than needed. The SmartBox system also provides positive end-row shutoff, eliminating the extra material often deposited at the end of each row. Because it is a totally enclosed or remotely operated system, there is little chance for operator exposure to the pesticide (*47*). Such precise metering provides accurate application of a new-generation soil insecticide, chlorethoxyfos, whose superior potency against pests allows rates one-eighth to one-ninth of older standard products.

Controlled Release Formulations. Controlling the release of pesticides through the use of polymeric barriers or other means is similar in concept to controlled-release medication—small amounts of chemical are released over a period of time. Historically, applications of a pesticide were usually in excess of what was needed for control in order to compensate for losses due to decomposition, volatilization, and other factors. This excess material not only increased costs but also increased potential risks to the environment. Formulating a pesticide to control its release reduces these risks and also offers several other potential advantages as listed in Table V (*48*).

Table V. Potential Benefits of Controlled-Release Formulations

- Increased stability of the active ingredient in the presence of additives or other pesticides
- Improved handling safety
- Reduced contamination of the food supply
- Prolonged residual activity
- Improved safety to crops
- Potentially lower application rates
- Reduced loss of active ingredient due to environmental factors
- Reduced environmental contamination
- Reduced odor

Adapted from ref. 48, p. 263.

Future Role of Pesticides in IPM

Pesticides will continue to play an important role in IPM programs. With reduced rates, increased safety, and reduced risk to the environment, pesticides can work hand-in-hand with all IPM control methods, delivering the added efficacy that will help sustain agriculture and support a growing population. In many ways, the use of pesticides in IPM is analogous to the use of medicine in human health care. In the past, medicines were often regarded as "miracle cures," prescribed with too little regard for chronic effects, undesirable side effects, and the potential contribution of other, holistic forms of treatment. Now, like pesticides, medicines are seen as an important part of the answer to human health care, to be used judiciously and in concert with other available therapies.

Acknowledgment

The author would like to thank Ann F. Birch, Editech, for her help in preparing this manuscript.

Literature Cited

1. Food and Agricultural Organization of the United Nations, *International Code of Conduct on the Distribution and Use of Pesticides* (Article 2), Rome, 1990.
2. Bottrell, D.R. *Integrated Pest Management*, Council on Environmental Quality, Washington, DC, 1979, pp 6-18.
3. Council for Agricultural Science and Technology (CAST), *Sustainable Agriculture and the 1995 Farm Bill*, Special Publication No. 18, Ames, IA, 1995.
4. Oerke, E-C.: Dehne, H.W.; Schonbeck, F.; and Weber, A., *Crop Production and Crop Protection*: Elsevier: Amsterdam, 1994: pp 5-8.
5. Hinrichsen, D. "Winning the Food Race," In *Population Reports*; Robey, B., Ed.; Population Information Program, Center for Communication Programs, The Johns Hopkins University School of Public Health: Baltimore, MD, 1997, Vol. XXV, No. 4.
6. Oerke, E-C.: Dehne, H.W.; Schonbeck, F.; and Weber, A., *Crop Production and Crop Protection*: Elsevier: Amsterdam, 1994: pp 34-37.
7. Bottrell, D.R. *Integrated Pest Management*, Council on Environmental Quality, Washington, DC, 1979, pp 1-5.
8. Flint, M.L.; van den Bosch, R. *Introduction to Integrated Pest Management.* Plenum Press: New York, 1981, pp 51-81.
9. Oerke, E-C.: Dehne, H.W.; Schonbeck, F.; and Weber, A., *Crop Production and Crop Protection*: Elsevier: Amsterdam, 1994: pp 52-71.
10. Anderson, W.P. *Weed Science: Principles and Applications*, 3rd ed.; West Publishing Company: Minneapolis/St. Paul, MN, 1996, p 59.
11. Plapp, Jr., F.W. "The Nature, Modes of Action, and Toxicity of Insecticides." In *CRC Handbook of Pest Management in Agriculture*; Pimentel, D., Ed.; CRC Press: Boca Raton, FL, 1981, Vol. II, p. 452.

12. Thompson, G.D.; Busacca, J.D.; Jantz, O.K.; Kirst, H.A.; Larson, L.L.; Sparks, T.C. "Spinosyns: An Overview of New Natural Insect Management Systems." *1995 Proc. Belt. Cotton Conf.*, Vol. 2, 1995, pp. 1039-1043.
13. Ragsdale, N.N.; Sisler, H.D. "The Nature, Modes of Action, and Toxicity of Fungicides." In *CRC Handbook of Pest Management in Agriculture*; Pimentel, D., Ed.; CRC Press: Boca Raton, FL, 1981, Vol. II, p. 471, 485, 486.
14. *New Partnerships for Sustainable Agriculture*; Thrupp, L.A., Ed.: World Resources Institute: Washington, DC, 1996.
15. Funderburk, J.; Higley, L; Buntin, G.D. "Concepts and Directions in Arthropod Pest Management." In *Advances in Agronomy*; Sparks, D.L., Ed.; Academic Press, Inc., San Diego, CA, 1993, Vol. 51.; pp 125-172.
16. Georghiou, G.P. "The Magnitude of the Resistance Problem." In *Pesticide Resistance: Strategies and Tactics for Management*, National Research Council, Committee on Strategies for the Management of Pesticide Resistant Pest Populations, Board on Agriculture, National Academy Press: Washington, DC, 1986; pp 14-43.
17. Roush, R.T.. "Management of Pesticide Resistance." In *CRC Handbook of Pest Management in Agriculture*; Pimentel, D., Ed.; CRC Press: Boca Raton, FL, 1981, Vol. II, pp 731-735.
18. Brown, L.R.; Flavin, C.; Postel, S. *Saving the Planet: How to Shape an Environmentally Sustainable Global Economy*, W.W. Norton & Company: New York, NY, 1991.
19. Benbrook, C.M. *Pest Management at the Crossroads*, Consumers Union: New York, NY, 1996.
20. Avery, D.T. *Saving the Planet with Pesticides and Plastic: The Environmental Triumph of High-Yield Farming*, Hudson Institute: Indianapolis, 1995.
21. Davies, J.E. "Health Effects of Pesticides." In *CRC Handbook of Pest Management in Agriculture*; Pimentel, D., Ed.; CRC Press: Boca Raton, FL, 1981, Vol. II, pp 439-444.
22. Wiles, R.; Davies, K.; Campbell, C. *Overexposed: Ogranophosphate Insecticides in Children's Food*, Environmental Working Group, Washington, DC, 1998.
23. Lamb, J.C., "Beware Junk Science," *The Washington Times*, January 8, 1997, A17.
24. Allen, W.A., Rajotte, E.G., Kazmierczak, R.F., Jr., Lambur, M.T., Norton, G.W. "The National Evaluation of Extension's Integrated Pest Management (IPM) Programs." VCES Publication 491-010, Virginia Cooperative Extension Service and USDA Extension Service, Blacksburg, VA, 1987; cited in Adoption of Integrated Pest Management in U.S. Agriculture, USDA, Agriculture Information Bulletin Number 707, 1994.
25. U.S. Department of Agriculture, Agricultural Research Service, "USDA Programs Related to Integrated Pest Management," USDA Program Aid 1506, 1993; ; cited in Adoption of Integrated Pest Management in U.S. Agriculture, USDA, Agriculture Information Bulletin Number 707, 1994.
26. *Adoption of Integrated Pest Management in U.S. Agriculture*, USDA, Agriculture Information Bulletin Number 707, 1994.

27 Peterson, L.G., Ruberson, J.R., Sprenkel, R.K., Weeks, J.R., Donahoe, M.C., Smith, R.H., Swart, J.S., Reid, D.J., Thompson, G.D., "Tracer Naturalyte Insect Control and IPM," *Down To Earth*, **1997**, Vol. 52, No. 1, p 29.

28. Congress of the United States, Office of Technology Assessment, "Risks and Regulations," Ch. 4, *Biologically Based Technologies for Pest Control*, Washington, D.C., 1995; pp 69-107.

29. Shani, A., "Integrated Pest Management Using Pheromones," *CHEMTECH*, Vol. 28, No. 3, **1998**; pp 30-35.

30. Congress of the United States, Office of Technology Assessment, "The Technologies," Ch. 3, *Biologically Based Technologies for Pest Control*, Washington, D.C., 1995; pp 31-68.

31. Greathead, D.J. "Natural Enemies in Combination with Pesticides for Integrated Pest Management." In *Novel Approaches to Integrated Pest Management*; Reuveni, R., Ed.; Lewis Publishers: Boca Raton, FL, 1995; pp 183-197.

32. Bottrell, D.R. *Integrated Pest Management*, Council on Environmental Quality, Washington, DC, 1979, pp 27-45.

33. Sailer, R.I. "Extent of Biological and Cultural Control of Insect Pests of Crops." In *CRC Handbook of Pest Management in Agriculture*; Pimentel, D., Ed.; CRC Press: Boca Raton, FL, 1981, Vol. II, pp 3-11.

34. Kogan, M., "Integrated Pest Management: Historical Perspectives and Contemporary Developments," *Ann. Rev. Entomol.*, Vol. 43, **1998**, pp. 243-270.

35. U.S. Department of Agriculture, Three Agency Release, Presidential Announcement Regarding IPM Adoption. Office of Communications, June 23, 1993.

36. Czapar, G.F., Curry, M.P., "Farmer Acceptance of Economic Thresholds for Weed Management," *Proceedings of the Third National IPM Symposium/Workshop*, Feb. 27- Mar. 1, 1996. (abstract)

37. Gonsulus, J.L., Buhler, D.D., "A Risk Management Perspective on Integrated Weed Management," *Journal of Crop Protection*, Vol. 2, **1998**, in press.

38. Norton, G.W., Mullen, J. *Economic Evaluation of Integrated Pest Management Programs: A Literature Review*, Virginia Polytechnic Institute and State University, Blacksburg, VA, Publication 448-120, 1994.

39. Mullen, J.D., Norton, G.W., Reaves, D.W., "Economic Analysis of Environmental Benefits of Integrated Pest Management," *J. Agric and App. Econ.*, Vol. 29, No. 2, **1997**, pp. 243-253.

40. Smith, E.G.; Knudson, R.B.; Taylor, C.R.; Penson, J.B. "Impacts of Chemical Use Reduction on Crop Yields and Costs," Agricultural and Food Policy Center, Dept. of Agric. Economics, Texas A&M University, Cooperating with Tennessee Valley Authority, College Station, TX, 1990.

41. Brown, H.M., Lichtner, F.T., Hutchison, J.M., Saladini, J.A., "The Impact of Sulfonylurea Herbicides in Cereal Crops," Brighton Crop Protection Conference - Weeds, 1995.

42. Criswell, J.T., Dunn, J., Cuperus, G., "Pesticide Use on Oklahoma Wheat between 1981 and 1995, *Proceedings of the Third National IPM Symposium/Workshop*, Feb. 27- Mar. 1, 1996. (abstract)
43. Sparks, T.C., Thompson, G.D., Larson, L.L., Kirst, H.A., Jantz, O.K., Worden, T.V., Hurtlein, M.B., Busacca, J.D., "Biological Characteristics of the Spinosyns: A New Naturally Derived Insect Control Agent," *Proc. Belt. Cotton Conf. Nat. Cotton Council of Am.*, San Antonio, TX, 1995, pp. 903-907.
44. Thompson, G.D., Busacca, J.D., Jantz, O.K., Borth, P.W., Nolting, S.P., Winkle, J.R., Gantz, R.L., Huckaba, R.M., Nead, B.A., Peterson, L.G., Porteous, D.J., Richardson, "Field Performance in Cotton of Spinosad: A New Naturally Derived Insect Control System, *Proc. Belt. Cotton Conf. Nat. Cotton Council of Am.*, San Antonio, TX, 1995, pp. 907-910.
45. Peterson, L.G., Porteous, D.J., Huckaba, R.M., Nead, B.A., Gantz, R.L., Richardson, J.M., Thompson, G.D., "Beneficial Insects: Their Role in cotton Pest Management Systems Founded on Tracer Naturalyte Insect Control, *Proc. Belt. Cotton Conf. Nat. Cotton Council of Am.*, Nashville, TN, 1996, pp. 872-874.
46. Moran, M.S., Inoue, Y., Barnes, E.M., "Opportunities and Limitations for Image-Based Remote Sensing in Precision Crop Management," *Remote Sens. Environ.*, Vol. 61, **1997**, 319-346.
47. DuPont Agricultural Products, Fortress Insecticide Product Literature, Wilmington, DE, 1995.
48. Tocker, S. "Evaluation of Controlled Release: Efficacy, Environment, and Commercial Aspects" In *Controlled Delivery of Crop-Protection Agents*; Wilkins, R.M., Ed.; Taylor & Francis: London, 1990; pp 261-277.

Chapter 14

The Economic, Health, and Environmental Benefits of Pesticide Use

Scott Rawlins

Commodity Policy and Program Specialist, Public Policy Division, American Farm Bureau Federation, Park Ridge, IL 60068

The debate over pesticide use is contentious because activist groups focus the general public's attention on pesticide risk only, ignoring information about why farmers use pesticides and the benefits pesticides provide to society. The result is an overly emotional debate that too often ignores broader scientific issues and the contributions of pesticides.

The debate over pesticide use has focused on several disparate issues. On one side there are concerns that pesticide use and the presence of pesticide residues on food and in water leads to widespread, adverse health effects. These concerns stem from studies done during a pesticide's registration process and that adverse effects from feeding pesticide chemicals to laboratory animals are transferred to humans with identical results, even at doses hundreds or thousands of times lower than doses that produce no adverse effects in laboratory animals. This is highlighted by the complete lack of empirical and consistent data and demonstrated examples where long-term exposure to minute traces of pesticide residues on food and through other exposures have ever caused any examples of adverse health effects. On the other side of the debate are farmers and others who use pesticides to reduce the risk of crop failure, who clearly see the economic and productive benefits of pesticide use. Resolving the differences of opinion in this debate has been difficult, leading to regulatory actions based on emotion rather than science.

The lack of a "middle ground" among these voices has lead to a polarization of the pesticide debate where consensus is impossible and meaningful debate is lost. Added to this is the tendency among activist groups to use the pesticide issue for highly emotional and alarmist messages to the general public, making it even more difficult to have meaningful debate.

As a result, science and economics take a backseat to other messages and issues. Without scientific or economic considerations, decisions regarding pesticide use and regulation tend to be skewed and don't reflect the benefits pesticides provide to society. Because of this, the decisions regarding pesticide use and regulation tend to be skewed and don't reflect the benefits pesticides provide to society. Because of this, the economic, health and environmental benefits provided by pesticides tends to get lost in the emotion of pesticide issues.

Undeniable Truths

Even though very few people agree on the issues related to pesticide use, there are several undeniable truths related to agriculture, food security and pesticide use that must be considered in the debate over pesticide risk versus benefits. These truths are:

- World population is expected to grow from 5.8 billion today to 10 billion by 2040. The challenge to feed 70% more people over the next 40 years presents significant challenges to world food production systems. How the world provides enough food for the world's burgeoning population without bringing new lands into production will be a major environmental challenge.
- Pests cause crop damage. Weeds compete for water, nutrients and sunlight. Insect damage to crops and fungal pathogens can cause major yield losses. Even farmers who use few or no "synthetic" pesticides recognize the need to control and manage various economical pests, by substituting synthetic pesticides with "organic" pesticides.
- Lower worldwide yields mean a need to cultivate more acres to feed a growing world population. In the United States, planted farm acreage equals roughly 330 million acres. A 10% reduction in per acre yields means it will take 11 acres to produce what we now produce on 10 acres. For the United States, a 10% yield reduction means 33 million more acres are needed to produce what we now produce on 330 million acres. The total land area of Arkansas equals about 33 million acres.
- Creating new farmland has environmental impacts. There are currently 28 million acres enrolled in the Conservation Reserve Program (CRP). If per acre yields drop, additional land will be needed to maintain production. The land most likely to be brought back into production in the U.S. is land from the CRP. Unfortunately, bringing environmentally sensitive land in back into production creates other environmental problems, including the destruction of wildlife habitat.
- New technologies will allow productivity to rise at historic rates. In 1963, the U.S. produced about 125 million bushels of apples on 500,000 acres. In 1998, the U.S. will produce almost 270 million bushels of apples on the same amount of land. Barring

economic disincentives, apple production is expected to rise at similar rates for the foreseeable future. Most of this yield increase is due to technological innovation, including pesticide use. The outlook for other crops is similar. Greater economic incentives in the form of higher commodity prices will bring about even greater increases in per acre productivity.

- There is still no substitute for food.

Economic Benefits Associated with Pesticide Use

Farm Bureau has sponsored two economic studies *(1)* to examine the economic impacts of reduced pesticide use. Both studies examined implications, trade-offs and yield and cost impacts associated with reduced pesticide use for domestic food production. The common thread between both studies is fairly simple and says that mandated reductions in pesticide use cause per acre yield reductions and increases in per unit production costs, given current technology and a lack of viable, effective and economic alternatives. While the extent and magnitude of these changes may be in disagreement, everyone agrees that forced reductions in pesticide without alternatives causes a decrease in food production, yields and quality.

To highlight the benefits of pesticide use, what would happen with a marginal change in the cost of food? To demonstrate and underscore the benefits of the U.S. agricultural economy, Farm Bureau recently established Food Checkout Day. Food Checkout Day is the day when the average American has earned enough income to pay for their entire annual food supply, including food consumed away from home. In 1998, Food Checkout Day was February 9.

In 1995, Dr. Robert Taylor of Auburn University examined the consumer and food safety tradeoffs from policies that reduce pesticide use on fruits and vegetables. *(2)* According to the study, the following impacts represent the effects of a 50 percent reduction in pesticide use:

- Domestic production would decrease by 6 percent;
- Exports would decrease by 10 percent;
- Imports would increase by 3 percent;
- Domestic consumption would decrease by 4 percent;
- Wholesale prices would increase by 17 percent; and
- Retail prices would increase by 9 percent.

Based on this information, delaying Food Checkout Day to March 9 means $49.8 billion (The U.S. spends $1.66 billion/day to pay for food) is diverted from other sectors of the economy to pay for food. Other sectors of the economy would then suffer, including cars, restaurants, and vacations. Pesticides allow Americans to spend money on items rather than on food.

Health Benefits from Pesticides

The message environmental activists send to the public is that pesticides cause serious health problems, especially to infants and children. If this is true, then the elimination of pesticides provide significant health benefits. What are they?

- Longer life spans?
- Reduced incidence of certain illnesses?
- Better health? Can benefits be quantified?
- Are their other benefits?

Activist groups can not quantify the health benefits of reducing or eliminating pesticide use. They can only claim that adverse health effects will be avoided, but they can't provide any examples of an adverse health effect from consuming minute traces of pesticide residues on food. Conversely, there is a growing body of evidence linking increased fruit and vegetable consumption and better diets with reduced incidence of cancer. In 1996, the National Research Council (NRC) released the report, "Carcinogens and Anticarcinogens in the Human Diet." The charge to the researchers who developed the report was to "examine the occurrence...and potential role of natural carcinogens in the causation of cancer (in humans), including relative risk comparisons with synthetic carcinogens and a consideration of anticarcinogens." The report draws several conclusions. Those conclusions include:

- "The great majority of individual naturally occurring and synthetic food chemicals are present in the human diet at levels so low that they are unlikely to pose an appreciable cancer risk."
- "Most naturally occurring minor dietary constituents occur at levels so low that any biologic effect, positive or negative, is unlikely. The synthetic chemicals in our diet are far less numerous than the natural and have been more thoroughly studied, monitored and regulated. Their potential biologic effect is lower."

In other words, natural chemicals present in the foods we eat pose a greater cancer risk than synthetic substances. However, both natural and synthetic substances are present in our diets "at levels so low that they are unlikely to pose an appreciable cancer risk."

The NRC report goes on to say:

- "Fruits and vegetables contain antioxidant vitamins that have demonstrated anticarcinogenic properties in laboratory studies; diets rich in fruits and vegetables are in fact associated with reduced cancer risks in humans."
- "In terms of cancer causation, current evidence suggests that the contribution of calories and fat outweighs that of all other individual food chemicals, both naturally occurring and synthetic."

In layman's terms, these conclusions sound a lot alike advice every parent dispenses -- watch your weight and eat a balanced diet that includes lots of fruits and vegetables. Except, 25 years ago, fruits and vegetables weren't always available year-round. They are now and we have incredible improvements in agricultural technology to thank for it. The anti-pesticide message sends consumers an entirely different message because it tells them to avoid fruits and vegetables because consuming them creates an imminent and certain hazard. Based on the NRC's findings, this represents irresponsible policy.

While some in our society fret over pesticides and the risk they may pose, other risk factors play a much larger role in the health of Americans. For example:

- According to the National Center for Health Statistics, 59% of American men and 49% of women are overweight. Ten years ago, 51% of men and 41% of women were this heavy. 73% of men and 64% of women over 50 are overweight. A record percentage of pre-teen children are now overweight.

Another concern in the pesticide debate is that we sufficiently protect infants and children from exposure to pesticide residues in food. In fact, the Food Quality Protection Act, the nation's new pesticide law passed on 1996, requires EPA to apply an additional tenfold safety factor for some pesticide products. This provision is the result of findings in the 1993 NRC report, *Pesticides in the Diets of Infants and Children.* While some dispute the findings of the NRC kid's report and suggest that children are not necessarily more sensitive to pesticide residues, the real reason behind including children in this debate is to use them as a shield to protect activist groups, politicians and EPA from criticism for broad-based pesticide cancellations and tolerance revocations. In some regards, children are now the central players in the pesticide regulatory debate and are being used to justify regulatory action. However, lost in this debate are consideration of the real health risks facing children.

USEPA Administrator Carol Browner often talks about kids and cancer and said during a recent speech that "we need to know more about whether environmental factors are in any way responsible for the alarming increase in new incidences of childhood cancer." The American Cancer Society (ACS) disagrees with her statement. First, the ACS would take issue that there is some sort of epidemic of childhood cancer. The ACS says "as a childhood disease, cancer is rare." Statistical evidence from the ACS also show there is no "alarming increase" in the incidence of childhood cancer as Browner states. There are 69.5 million children in the U.S. under the age of 18. The ACS estimates 8,800 new childhood cancer cases in 1997, an incidence rate of 12.6 per 100,000 population. The incidence rate per 100,000 children for the last ten years is relatively flat. Variations in the rate over this same period are up some years and down in others. The number of new cases is increasing due to increases in the child population and better detection methods, but the number of new cases per 100,000 children is not increasing. So, Browner's assertion that there is an "alarming increase in the incidence of childhood cancer" is not true. Plus, according to the ACS, one-third of all new cases is from leukemia. The ACS states that the causes of "most leukemias are unknown" and that "persons with Down's Syndrome and certain other genetic abnormalities have higher than usual incidence rates of leukemia." Browner wants to lay the blame on environmental factors, but she does so without any evidence. While cancer is tragic, especially among children, misusing information about children to gain political advantage is irresponsible because it misdirects resources and attention away from the real causes.

You don't have to look very hard to find the real problems affecting kids. Just pick up the newspaper or watch the network news to find the real kid issues we should be worrying about.

- The September, 1997 issue of the journal "Pediatrics" reports that only 1% of American young people ages 2 to 19 eat healthy diets.
- The National Cancer Institute (NCI) in a 1997 study published in the "Archives of Pediatrics and Adolescent Medicine" says that when kids do eat their vegetables, it's likely to be french fries. Susan M. Krebs Smith, a nutrition researcher with the NCI says that "grow-ups are not eating their vegetables, and kids aren't either."
- U.S. Department of Agriculture food consumption data *(3)* demonstrate that low income kids eat far fewer fruits and vegetables than higher income kids. As income rises, so does fruit and vegetable consumption.
- Accidents (most from motor vehicles) and deaths from gunfire remain the first and second leading cause of death among kids by far. Gunfire is the leading cause of death among black males age 15-19 with a death rate of 153.1 per 100,000, more than five times higher than the death rate among white males of the same age.
- According to the National Child Abuse Coalition, 3.1 million children were reported as abused or neglected in 1995. About 10,000 kids are killed each year by child abusers. Overall, the total number of reports of child abuse and neglect nationwide increased 49% since 1986. The U.S. Department of Health and Human Services (HHS) reports similar numbers. HHS says "the total number of child abuse and neglect cases rose from an estimated 1.4 million in 1986 to an estimated 2.8 million in 1993," a 100% increase.
- In October of 1997, the World Bank sponsored a meeting on looming food shortages. The International Food Policy Research Institute warns that 2 out of 5 children in southeast Asia will be without enough to eat by 2020. Forty million children in Africa will be malnourished.

Perhaps most important are the health benefits pesticides supply to low income Americans, consumers who already consume 20 percent less fresh fruits and vegetables than higher income consumers. By making fruits and vegetables less expensive, pesticides assist low income Americans in their efforts to consume a healthier and more balanced diet. A growing body of literature supports the concept that wealthier individuals are healthier, primarily because they can afford items that reduce risk. The corollary, of course, is that poorer individuals are less healthy. *(4)*

Pesticide scare messages have a psychological negative component that serves to frighten consumers by implying they should make other diet choices because of the possible presence of a substance that causes tumors in laboratory animals at high doses. Based on the NRC *Anticarcinogen* report, sound public health policy with the goal of reducing overall cancer risk, should be grounded in policies that encourage fruit and vegetable consumption.

Environmental Benefits Associated with Pesticide Use

The most compelling reason of the benefits from pesticide use are described by Nobel laureate Norman Borlaug:

> "...[I]f U.S. farmers used the agricultural technology of the 1930s and 1940s to produce the harvest of 1985, they would have to convert 75% of the permanent pasture lands in the U.S. or 60% of the American forests and woodland areas to cropland. Even this may be an underestimation, since the pasture and forestlands are potentially less productive than the land now planted to crops. This would greatly accelerate soil erosion and destroy wildlife habitats and recreational areas."

The benefits of pesticides accrue to all of society, not just to farmers, and their consideration in pesticide regulatory decisions is critical for a reasoned and coordinated policy. The benefits of pesticide use must be balanced with risks along with the need to feed a world population that is growing by nearly 100 million people every year.

A risk-only approach to pesticide regulation does not reflect the contribution of pesticides to our food supply. It is important to note the benefits society derives from the safe and judicious use of pesticides. Research from Texas A&M University *(5)* catalogs the benefits derived from pesticide use. These benefits include:

- Pesticides reduce the risks of crop failure and stabilize food production.
- Pesticides increase yields and allow food to be produced on less land. Land that would otherwise be needed for food production can be devoted to wildlife habitat and other beneficial uses. Pesticides also allow environmentally fragile lands to be idled. Fewer farmed acres reduces the amount of water needed for irrigation.
- Pesticides prevent soil erosion resulting from increased cultivation to control weeds.

- Pesticides reduce farm costs. Reduced costs allow us to compete in world markets. Lower farm costs also translate to lower food costs which encourage consumption of foods important to health.

- Pesticides allow food to be grown domestically, rather than depending on imports where we have little to no control over food production methods.
- Pesticides improve the quality and storability of food. Consumers can expect more perishability at the marketplace as a result of pest infestation and consumer rejection of products with poor appearance and quality if farmers are forced to arbitrarily reduce pesticide use. Consumers can expect poor quality foods if they are typically stored for long periods. High quality foods are essential for meeting export standards as well. Customer countries will reject U.S. products if they do not meet quality or phytosanitary standards.
- Pesticides decrease farm labor requirements. History has shown that it is difficult to attract labor to agriculture due to the often difficult working conditions.

Perhaps the greatest environmental benefit supplied by pesticides is that pesticides allow food to be produced on fewer acres. In 1994, the University of California-Davis, published the results from a multi-year study examining and comparing various production systems. The study looked a quality, cost and yield impacts from three production systems: conventional, organic and a low input system. When comparing conventional yields with organic yields, the study found that for all crops, except beans, organic yields were always lower than conventional yields. For example, organic tomatoes needed 30% more acres to produce the same total output as conventional tomatoes. Organic safflower needed 65% more acres to produce the same total output as conventional safflower.

Conclusions

Pesticides provide significant benefits to society. They provide Americans with a year-round supply of fresh fruits and vegetables. They are a risk management tool for farmers. They allow consumers to use their income elsewhere. Plus, they feed a growing world with minimal environmental impact.

There is a lot of good news for consumers. The supply of food is bountiful, quality is unparalleled, variety is ever-expanding and prices are reasonable. The American farmer/government/university food production system is unrivaled. Our quality of life and health provide sufficient evidence and argument to build upon our current system.

It is important to note that while modern technology has greatly improved our ability to measure or detect the tiniest trace of chemicals in food, we have had no increase in our ability to make these numbers useful or meaningful to the food policy process. This results in periodic food safety scares. This does not mean that our current system is broken and in need of an overhaul. Rather, it suggests the need to carefully change pesticide policy to reflect scientific advancement. Those who argue that pesticides provide no benefit to society are ignoring scientific evolution and modern technology in an attempt to divert our attention to insignificant risks away from the risks we can change.

Literature Cited

(1) Impacts of Chemical Use Reduction on Crop Yields and Costs (Knutson, Smith, Penson and Taylor) & Economic Impacts of Reduced Pesticide Use on Fruits and Vegetables (Knutson, Hall, Smith, Cotner, Miller)

(2) C. Robert Taylor, "Economic Impacts and Environmental and Food Safety Tradeoffs of Pesticide Use Reduction on Fruits and Vegetables."

(3) Steven M. Lutz, David M. Smallwood, James R. Blaylock, Mary Y. Hama, Changes in Food Consumption and Expenditures in Low-Income American Households During the 1980's, USDA/ERS Human Nutrition Information Service, 1993

(4) "Mortality Risks Induced by Economic Expenditures," Risk Analysis, Vol. 10, No. 1, 1990, p. 147-159.

(5) Ronald D. Knutson, Charles R. Hall, Edward G. Smith, Samuel D. Cotner, John W. Miller, "Economic Impacts of Reduced Pesticide Use on Fruits and Vegetables," 1993

Chapter 15

Beneficial Impacts of Pesticide Use for Consumers

Leonard P. Gianessi

National Center for Food and Agricultural Policy, 1616 P Street, N.W., Washington, DC 20036

Widespread use of synthetic organic pesticides has contributed enormous benefits to U.S. consumers. Chemical pest control has contributed to dramatic increases in yields for most major fruit and vegetable crops. As a result, U.S. consumers have an inexpensive year-round supply of a wide variety of foods. Chemical pesticides have contributed, as well, to the appearance and quality of produce – traits highly prized by U.S. consumers.

The benefits of chemical pesticides for consumers can be seen in terms of: (1) the price of organic foods (on average 57% higher), and (2) the consequences of bans on pesticide uses (a decline in production of certain apple varieties following the Alar ban).

The beneficial impacts of pesticide use for consumers fall into four categories:

- Inexpensive food
- Plentiful, year-round supplies of food
- A wide variety of fruits and vegetables
- High quality/appearance of foods

Many of the benefits of pesticide use for consumers result from enormous increases in crop yields. High yields are one of the reasons that food prices are low in the U.S. and food is plentiful on a year-round basis. A succinct summary statement regarding the role of pesticides in increasing crop yields with subsequent benefits for consumers was provided by the National Research Council in a 1993 report (*1*):

> Chemical pest control has contributed to dramatic increases in yields for most major fruit and vegetable crops. Its use has led to substantial improvements over the past 40 years in the quantity and variety of the U.S. diet and thus in the health of the public.

Food Production Increases

Numerous examples of dramatic improvements in crop yields can be cited to support the beneficial role of pesticides.

In California's strawberry acreage, average yields were 5 to 10 tons per acre in the 1950's, even though the yield potential of new cultivars was 20 to 30 tons per acre (2). California strawberry growers began fumigating their acreage with methyl bromide in the mid-1960's and strawberry yields have averaged 20 to 25 tons per acre for the past 30 years. Research demonstrated that soil-borne diseases had been claiming the greater part of the untaken harvest, and the chemical solved the problem.

In Maine, wild blueberry fields have been harvested for hundreds of years. Weeds were a serious problem, that was controlled only recently. In the mid-1980's an herbicide (hexazinone) was made available for Maine blueberries, and production doubled as a result – from 20 million lb/yr to 40 million lb/yr (*3*).

Production from Florida serves as a winter source for many fruits and vegetables. One major crop from Florida in the winter is sweet corn, that is delivered fresh to most cities on the East Coast. Florida is the major fresh corn producing state with 500 million pounds in annual production (*4*). The fresh corn industry did not exist in Florida before the introduction of synthetic chemicals in the late 1940's. Prior to the development of synthetic insecticides, it was not even possible to grow the crop in Florida because of the many insect problems.

Apple production certainly has benefited from chemical pesticide use. There are numerous insect and disease problems of apples, including codling moth, apple maggot, apple scab and bitter rot. These pest problems have plagued apple growers since the beginning of widespread commercial production in the early 1900's. Thus, commercial apple trees were sprayed regularly with inorganic compounds such as arsenic and sulfur from 1900 to 1950 (*5*). When the switch-over to synthetic chemicals occurred in the early 1950's, there were tremendous production increases in apples. In many states, per tree apple production increased by 100% to 200%: New York (110%), Pennsylvania (157%), Virginia (209%) and Michigan (109%). Not only did the synthetic chemicals do a better job of controlling the pests, but they also were less harmful to the trees. The arsenic and lime sulfur, while controlling pests, actually damaged the trees and reduced their production.

Potato production has been plagued with numerous diseases and insects. One of the most serious disease problems of potatoes is late blight – the disease that led to the Irish potato famine. Once again, consumers have benefited from the use of modern chemicals in potato production. For example, most potato growers in Maine switched from using copper to using the EBDC fungicides when they became available in the early 1950's. Average potato yields in Maine increased from 17,000 lb/A in the early 1940's to 26,000 lb/A in the early 1950's (*6*).

Organic Production

One way to evaluate the value of synthetic chemical pesticides to growers and consumers is to examine the economics of organic food production. Sales of organically produced food and fiber have increased steadily in the last few years. In 1995, organic sales totaled $2.8 billion. Organic growers do not use synthetic chemical pesticides. However, in general, organic foods cost the consumer more. *Consumer Reports* recently conducted a comparison shopping test and found that, on average, organic foods cost 57 % more than conventionally produced foods (*7*). One of the reasons that organic foods cost more is the use of less effective, more costly pest control techniques in place of synthetic chemical controls. As a result of using these less effective, more costly techniques, organic growers need to receive a higher price per unit of output in order to make a profit.

One of the really costly operations in organic production is weed control without the use of synthetic chemicals. Vegetable crops like lettuce, tomatoes, carrots, onions and celery can be overwhelmed by weeds that cause yield losses by competing for soil moisture, light and space. Weeds also interfere with harvesting. Commercial growers apply synthetic herbicides to fields, often before emergence of weeds and gain season-long control of germinating weeds, usually at a cost of $55 to $60 per acre for vegetable crops (*8*).

Organic growers use hand laborers with hoes to remove weeds from fields. Detailed cost of production budgets from the University of California indicate that an acre of lettuce or garlic requires 18 hours of hand labor for weeding at a cost of $150/A. Crops like onions require more than 70 hours of hand weeding at a cost of $600/A (*9*). In arid California many vegetable crops are irrigated with drip irrigation pipes right down the row of plants and weed growth is controlled as well with irrigation. Weeds are a problem in California, but not as serious a problem as in states with normal rainfall – such as New Jersey. Weed control experiments in New Jersey indicate that 200 and 1000 hours of weeding labor are needed for lettuce and onions, respectively, to produce yields equivalent to acreage treated with herbicides (*10*). The high costs of hand weeding result in higher costs and prices for organic food and fiber. Organic cotton is being produced in a few arid regions, in the southwest (Texas, California). A California organic cotton grower was quoted in the *New York Times* to the effect that lower yields (20% lower) and higher labor costs for hoeing were two of the main reasons that organic cotton sells to the consumer at three to four times the cost of conventional cotton (*11*).

Retrospective Analysis

Another way to determine the benefits of pesticide use for consumers is to examine what happens to food prices and supplies when pesticides are banned. This is a contentious issue with many activist groups claiming that, historically, no price or quantity effects have occurred when pesticides have been banned. The Environmental Working Group claims that the cancellation by EPA of 200 uses of 12 pesticides since 1985 had absolutely no effect on the price or the availability of food anywhere in the U.S. (*12*). Although it may appear on the surface that there are plenty of fruits and vegetables at an affordable price, past bans on pesticides have led to a measurable effect on supply and price. These effects are subtle, difficult to measure, and generally specific to certain types of products or regions of the country.

Cranberries are used widely in many consumer drinks and also are sold as dried and fresh products. Cranberries are subject to many different diseases, such as black rot. Between 1988 and 1993, U.S. cranberry growers lost the use of six fungicides that had been used for control of diseases of cranberries, including captafol, the most effective fungicide. These cancellations were not the result of EPA rule making, but rather resulted from voluntary cancellations by registrants. One effect of dropped fungicide registrations for cranberries has been a major decline in the production of cranberries for the fresh market. In Massachusetts, utilization of cranberries for the fresh market declined by 50% between 1983-1993 (*13*). Growers and shippers cannot take the risk of shipping cranberries in their fresh form with the possibility that they are diseased. The rots might appear in grocery stores or households. Thus, consumers who prefer fresh cranberries are having a difficult time finding them as a result of the lost fungicide registrations.

The discontinuation of the use of Alar in apples produced effects that were variety specific. Before the ban, the two varieties of apples that relied most heavily on Alar were Stayman and McIntosh. About 40% of the nation's acreage of McIntosh and Stayman were treated with Alar before the ban (*14*). Stayman apples are prone to develop serious cracks while still on the tree. This is a physiological disorder that Alar applications control. Following the ban on Alar, Stayman apple growers had no effective way of preventing the cracking. As a result, Stayman trees were pulled out of the ground. In Pennsylvania, apple tree censuses in 1986 and 1993 indicated that 30,000 Stayman trees were removed *(15)*. Consumers who prefer the Stayman variety of apples have not been able to enjoy this variety as before. McIntosh apples are prone to two physiological disorders that Alar prevents: 1) They tend to fall off the trees before maturity, and 2) They frequently lack adequate fruit color. Following the ban on Alar, a large number of the McIntosh apples fell off the trees before maturity and were not picked up for the fresh market. A large number of McIntosh apples did not color adequately and were not marketed fresh. As a result, New York State production of McIntosh apples for the fresh market declined 18%, while the wholesale price rose by 6% (*16*). Consumers who prefer fresh market McIntosh apples have not been able to find them as easily.

Food Quality

American consumers have high quality expectations for fruits and vegetables. The fruits and vegetables must be attractive in appearance and free from insect or other damage. Several groups have suggested that these marketing standards should be changed and consumers should accept surface blemishes and scars that would allow growers to reduce pesticide use (*17*). Some examples of surface blemishes include apples that are damaged by shallow feeding of the red banded leaf roller or small areas on the apple surface with dried apple scab lesions. A study was conducted at Michigan State University to determine what amount of apple surface damage consumers are willing to accept (*18*). The results suggest that consumers are willing to accept only a minor amount of pest damage in order to obtain reductions in pesticide residues. There is a three cent price penalty for each 1% increase in surface area damage. Focus groups indicated that worm damage is unacceptable to consumers.

For processed foods, some analysts have suggested that a different standard for pest damage ought to apply. For example, David Pimentel of Cornell University argues that for processing tomatoes the tomato fruitworm is a cosmetic pest (*19*). After all, the tomatoes are ground up into paste and sauces and insect fragments are hard to detect. The individual tomatoes attacked by the fruit worm are contaminated with worms and worm frass. Currently, the percentage of worm infested tomatoes acceptable for processing is very low. Recently, the California Processing Tomato Advisory Board considered a proposal to allow more insect fragments in processed tomato shipments. However, it was believed that consumers would prefer as few insect fragments as possible in processed tomatoes. The result was that there was no change in the allowable amount of insect damage in processed tomato shipments (*20*).

Summary

The widespread use of synthetic pesticides has produced enormous benefits for U.S. consumers. Fresh fruit and vegetables are inexpensive and available on a year round basis. The appearance and quality of produce is maintained to high standards because of the effectiveness of chemicals in controlling diseases and insects. The introduction of synthetic chemicals in the late 1940's and early 1950's led to a tremendous yield increase in U.S. food production and has made it possible for most foods to be sold at a low price.

Literature Cited

1. National Research Council. *Pesticides in the Diets of Infants and Children*; National Academy Press: Washington, DC, 1993, Chapter 1; pp 13.
2. Wilhelm, S; Paulus, A.O. *Plant Dis.* **1980**, *64*, 264-270.
3. Yarborough, D.E.; Bhlowmik, P.C. *Acta Hort.* **1989**, *241*, 347.
4. USDA. *Agricultural Statistics 1995-1996*; U.S. Government Printing Office: Washington, DC, 1996, Chapter IV; p 11.
5. Marlatt, C.L. *Yearbook of the Department of Agriculture* **1904**, pp 461-474.
6. USDA. *Potatoes*; Bureau of Agricultural Economics, Statistical Bulletin No. 122; Government Printing Office: Washington, DC, 1983; pp 30.
7. *Consumer Reports* **1998**, *January*, 12-18.
8. Hinson, R; Boudreaux, J. *Projected Costs for Selected Louisiana Vegetable Crops, 1996 Season*; A.E.A. Information Series No. 142, Louisiana State University: 1996.
9. Klonsky, K.; Tourte, L.; Chaney, D.; Livingston, P.; Smith, R. *Cultural Practices and Sample Costs for Organic Vegetable Production in the Central Coast of California*; Giannini Foundation Information Service Series No. 92-4, University of California: 1994; pp 28, 33, 37, 41, 45.
10. Majek, B.A. *Proceedings of the Thirty-ninth Annual Meeting of the Northeastern Weed Science Society* **1985**, *39*, 124-125.
11. Schneider, P. *The New York Times* **June 20, 1993**, 1,11.
12. Environmental Working Group. *Pesticide Industry Propaganda: The Real Story*; Washington, DC, 1995; pp 8-9.
13. USDA. *Noncitrus Fruits and Nuts, 1994 Summary*; U.S. Government Printing Office: Washington, DC, 1995.
14. USEPA. *The Economic Impacts and Social Benefits of Daminozide Use on Apples: An Empirical Study*, April 1985.
15. USDA/Pennsylvania Dept. of Agriculture. *1992 Pennsylvania Orchard & Vineyard Survey*; 1993; p9.

16. USDA/New York State Department of Agriculture and Markets. *Marketing New York State Apples, 1993 Crop.*
17. Rosenblum, G. *On the Way to Market: Roadblocks to Reducing Pesticide Use on Produce*; Public Voice for Food and Health Policy: Washington, DC, 1991; pp 34-46.
18. van Ravenswaay, E.; Hoehn, J. *Willingness to pay for reducing pesticide residues in food: Results of a nationwide survey*; Staff Paper No. 91-18, Michigan State University: East Lansing, 1991.
19. Pimentel, D.; Kirby, C.; Shroff, A. In *The Pesticide Question Environment, Economics, and Ethics*; Pimentel, D.; Lehman, H., Ed.; Chapman & Hall: New York, NY, 1993; pp 85-105.
20. Zalom, F. G.; Jones, A. *J. Econ. Entom.* **1994**, *87*, 181-186.

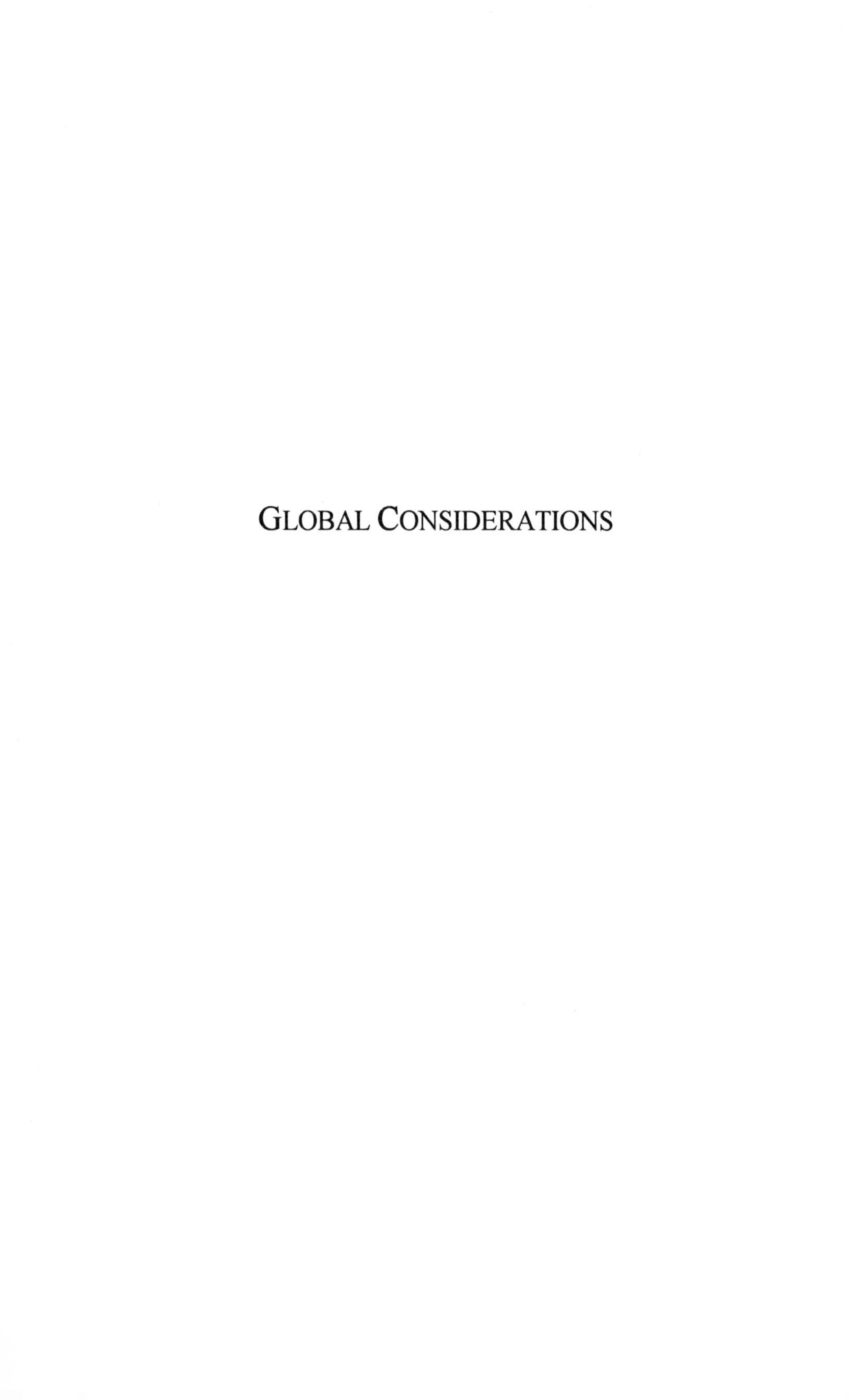

Global Considerations

Chapter 16

Global Harmonization of Pesticide Registrations

K. S. Rao

Global Risk Assessment Leader, Dow AgroSciences,
9330 Zionsville Road, Indianapolis, IN 46268-1054

The need for global harmonization in testing and assessment for health and safety of pesticides has been recognized for many years. Due to budgetary constraints on governments, harmonization has become a necessity on the part of governments and the industry alike. Unfortunately, many different testing and evaluation guidelines have been adopted by various countries, creating significant differences in registerability criteria. This recognition prompted the Organization for Economic Cooperation and Development (OECD) to create a platform known as the Pesticide Forum which generated recommendations for the harmonization of data requirements, guidelines as to methodologies for generating data, and a common format for submitting data (Dossier) by the industry and a uniform process for evaluating and reporting summaries by the government agencies (Monograph), among others. These emerging processes are still in the evolutionary stages and have not been fully adopted by various governments.

The goal of harmonization is to share expertise and pool resources to build the capacity and capabilities necessary internationally for the sound registration of pesticides to protect human health and the environment. The benefits can be potentially great – improving food safety, reducing huge regulatory and resource burdens on national governments and corporations, improving science, increasing information exchange and reducing trade problems. The use of pesticides to control pests and diseases is important for the production of sufficient and sustainable quantities of safe and affordable food (GIFAP, 1984). However, the use of these agents can sometimes leave residues (the pesticide or its degradates) in or on plant parts used as human food or animal feed commodities which enter into international commerce.

The lack of harmonized standards for registration of pesticides around the world has resulted in the establishment of inconsistencies (or lack of) residue limits of pesticides in international commerce. One recent case involved procymidone, a fungicide used in Europe to treat wine grapes but unregistered in the U.S. The lack of a U.S. import tolerance for procymidone resulted in detention of European wine and loss of millions of dollars to growers and processors in Europe. Until governments adopt a greater commitment to international harmonization, the current inconsistency in maximum residue limits (MRL's) seems likely to continue.

The export of crops can grow ever more difficult if governments continue to add a mountain of unharmonized national trading standards. Commodities complying with standards of one country risk rejection in another, even when the residues present are of no significant risk to health. Incomplete harmonization of MRL's can result in non-tariff trade barriers.

To market or ship a product internationally, companies must cope with different regulatory systems and attempt to develop labels and material safety data sheets (MSDS) to satisfy the varying requirements. These different requirements may constitute a technical barrier to trade. These barriers to participation in international trade could be effectively eliminated by a globally harmonized system, and the costs of compliance with varying international requirements can be significantly reduced. If all systems use the same criteria and acceptable methodologies, there will be no need to test the same chemical several times for compliance with the differing requirements of the various systems.

The need for international pesticides harmonization has been recognized for many years. Many governments are beginning to realize the difficulties of regulating pesticides on a strictly national basis. Preliminary moves to encourage harmonization of international pesticide control mechanisms are in motion. This important development can prove beneficial in resolving many of our difficult and expensive problems with duplicative non-harmonized registration requirements around the world.

Differences in the data requirements among various countries and in the methodologies accepted by them for the generation of data result in substantial additional costs for industry to address issues that often already have been investigated.

Pesticide manufacturers and regulators can benefit from better harmonized pesticide registration systems. National governments generally rely on own independent review instead of taking advantage of the work of other governments and international organizations. Various government agencies are spending considerable time and effort independently reviewing essentially the same data. Harmonization has the potential to greatly benefit all regulatory bodies by reducing the amount of duplicative review.

Some processes in place and some recent developments in regional and international harmonization activities that are in various stages of formation and development are summarized in the following paragraphs.

Codex Alimetarius Commission

One of the earliest attempts to harmonize food safety standards is carried out through the Codex Alimentarius Commission (CODEX), which is a subsidiary body of the World Health Organization (WHO) and the Food and Agricultural Organization (FAO). This organization has designed programs to protect the health of the consumer for more than 35 years through the CODEX process (Frawley, 1987). With representatives from 124 countries, CODEX is an active organization that provides an international forum for the discussion of issues related to food safety standards. The World Health Organization (WHO) includes the Joint Meeting of Pesticides Residues (JMPR) to consider health aspects, with the Food and Agricultural Organization considering use patterns and residues. The FAO panel proposes MRL's which cover a full spectrum of worldwide usage (Maybury, 1989). Since 1961, experts from governments and academia have evaluated over 230 pesticides. Monographs are issued by the JMPR pesticide reviews of studies relevant to the establishment of MRL's of pesticide residues in foods.

The sound and practical solution lies in all countries adopting the Codex values now. The pesticide industry therefore calls upon the countries for the adoption of additional MRL's or increased MRL values, to take full account of the Codex recommendations. Outside the United Nations system, there is no internationally recognized process of MRL harmonization that covers the whole world. The overwhelming majority of JMPR recommendations can safely be adopted without compromise to human health. However, because of different national guidelines and criteria national authorities and will often argue for different MRLs (Wessel, 1992).

The European Union

One of the more advanced trading groups is the European Union (EU). A well organization infrastructure has allowed the EU to advance considerably in the harmonization effort. The registration process in the EU is being harmonized under Directive 91/414 which has been in effect since 1993 (EC, 1993). Under the new directive, Member States have a common set of data requirements, study guidelines and decision-making criteria. However, even in this highly "harmonized" system there is no EU-wide registration per se. Only an active ingredient is authorized on an EU basis; the end-use products must be registered by individual Member States. Even in the EU, the active ingredient authorization is a consultative process that can be influenced by the policies and political agenda of the individual Member States.

Organization For Economic Cooperation And Development (OECD)

With respect to chemical safety since the 1970's major harmonization efforts have been made through committees of the OECD. However, the urgent need for global harmonization in the field of pesticides in the late 1980's has prompted the recent development of the Pesticide Forum Project organized under the auspices of the OECD to oversee the work of the pesticide program. In the formation meeting held in Saltsjobaden, Sweden on October 29-31, it was concluded that the OECD would be the lead organization for development of internationally harmonized test guidelines for pesticide registration by all member countries. With regard to data requirements, the OECD will develop a common core data sets to be mutually acceptable for registration of pesticidal active ingredients. This effort by the OECD will provide a workable process by which the differences in test guidelines and data requirements can be resolved for moving towards global harmonization. The long-term goal will be to develop common procedures for conducting data reviews and characterizing hazards so that OECD member countries can share the re-registration burden by agreeing to accept re-registration reviews among member Countries.

As a result of the various recommendations coming out of the Pesticide Forum, the OECD secretariat has initially undertaken the following projects:

1. Data Requirements: The ultimate goal of work in this area is to increase the ability of OECD countries to accept data and data reviews among member governments making it possible to share the burden of registration and re-registration of pesticides. Greater international harmonization of pesticide data requirements can also reduce the need for duplicative testing by industry, thereby saving resources, eliminating excessive animal studies, and easing barriers to trade.

   The OECD secretariat has collected from member states an inventory of current national data requirements for registering pesticides. The objective of this project is to:
   - identify the full set of data requirements imposed by OECD member countries for pesticide registration;
   - compare similarities and differences among the requirements to develop an agreed-upon common set of registration data requirements. Countries could use this common set but would not be prohibited from requiring additional data on a case-by-case basis.

In general, the survey results showed that in many test areas there is a significant amount of similarity in pesticide data requirements. The greatest similarity occurs in chemical identity, physical-chemical properties, toxicology, efficacy, mode of action, handling and analytical methods. The greatest differences occur in data requirements in the area of environmental fate, ecotoxicity, occupational exposure, and biological pesticides.

The survey also revealed a substantial overlap among member countries in the data elements required for registration. In practice, countries generally have the authority to waive a requirement if it appears not to be relevant. It should also be noted that in every test area, countries added further data elements to those listed in the questionnaire. In some cases the added data appear to be required to test for the same end point as data elements listed in the questionnaire, but with differences in test details such as duration, test species, etc. Finally, the survey results show that all responding countries had established data requirements for new pesticide registrations and that most had programs for re-registrations of pesticides.

2. Test Guidelines: OECD Member countries desire a common approach for the testing of pesticides which can be achieved through development of more standardized tests. The OECD secretariat was asked to conduct a survey of Member countries views on which existing test guidelines relevant to pesticides need re-examination and which new guidelines should be developed. The continued development and revision of OECD Test Guidelines to pesticides will remain a high priority activity. Nearly 100 guidelines are in various stages of preparation, review, and finalization.

3. Pilot Project: A pilot project was initiated to compare, for a selected number of pesticides, the re-registration data reviews done by OECD member countries or international organizations. The objectives of this pilot project are to:
   - collect general information on pesticide review processes in member countries;
   - gain an appreciation of the type of data reviews produced;
   - compare documents reviewing the same study or general test area in order to determine where major similarities and differences in scientific review may occur, to identify the basis of any significant differences; and thereby
   - recommend future work to improve international cooperation and harmonization in pesticide re-registration.

The pilot project has selected seven pesticides for which member countries indicated that recent reviews are available. They are amitraz; dinocap; dicofol; diazinon; iprodione; endosulfan; and pyridate. Germany, Sweden, Denmark, Switzerland, Finland, The Netherlands and the United States took the lead roles in comparing the data reviews on these chemicals. The role of the lead country was to compare data reviews submitted by participating countries and draft a report compiling similarities and differences in review process and data reviews with comments.

This pilot project revealed that, although subtle differences existed among countries in data reviews, these differences did not contribute to the overall conclusions or the resulting registration.

4. Hazard/Risk Assessment

The work on hazard/risk assessment is done by the Hazard Assessment Advisory Body which focuses on environmental and worker exposure assessment.

Based on the progress made so far, the Pesticide Forum has proposed that the activities are broadened to reflect:

- the important work on the harmonization of the format of a) industry data submissions for registration, and b) country data review (assessment) reports

- the shift in focus from old pesticides to all pesticides in the work sharing of pesticide assessments; and
- the expansion of the work on risk reduction beyond information sharing to include the development of ways to measure progress with risk reduction.

5. Registration Harmonization

The OECD has led the way in providing a forum for industrialized countries to identify common problems associated with regulating pesticides and to develop strategies for addressing them. There are two main projects in registration harmonization; the exchange of data reviews between governments, and the harmonization of submissions and reviews.

- Exchange of reviews

Regarding the exchange of data reviews, any OECD member country regulatory authority can request through the OECD secretariat an existing review of a study in support of registration. Practically every OECD country has exchanged such reviews on various pesticides. Once a country receives the review from another country, it is not clear what for they use the review to what type of decisions they make. The need to protect proprietary rights and Confidential Business Information (CBI) during the exchange of data review reports is paramount to the pesticide industry. Industry has expressed concern regarding confidentiality of data and the reviews and proprietary protection of data within countries.

- Dossiers/Monographs

Regarding the harmonization of submissions and reviews, the Pesticide Forum of the OECD in 1994 placed high priority on the harmonization of country data review reports and industry dossiers. It was agreed that the European Union (EU) formats being developed for these documents should be used as a starting point for development of more detailed guidance documents that could be adapted by all OECD Member countries.
A three-phased approach was agreed:

- Familiarization with the EU formats (especially by non-European countries);
- development of more detailed guidance to supplement the EU format, regarding what countries should include in their data review reports in particular study areas; and
- development of OECD formats for both country data review reports and industry dossiers.

A series of meetings in Dublin, Ireland during 1997-98 involving a sub-committee of the Pesticide Forum with industry involvement resulted in the publication of a framework which outlines the preparation of industry data submissions (dossiers) and government regulatory data review report summaries (monographs). The new formats are not intended to force OECD countries to make similar regulator decisions. Rather, their purpose is to prevent industry and government from having to do redundant work to define to the extent possible a common data set, and to encourage mutual acceptance of government reviews. The latter is modeled after the Joint Meeting of Pesticides (JMP) United Nation Codex System of evaluations which was incorporated into the World Trade Organization/General Agreement on Trade and Tariffs. Adopting this system can reduce redundancies among governments in their reviews and retain the high quality of currently done within OECD countries. The degree to which individual countries use a harmonized OECD monograph to make decisions will have to evolve over time as regulators gain confidence in the process.

6. Risk Reduction

The Pesticide Forum has also taken on cooperative work to develop measurements of the level of risk reduction achieved by pesticide regulation. The OECD is engaged in two major programs in the area of risk reduction. The first is a measurement of pesticide use and related changes in use to affect changes in risk. With regard to this, the workshop noted that pesticide registration offers a foundation for risk reduction - by providing for the evaluation and control of risks associated with individual pesticide uses - but that a wider approach, which addresses risks more comprehensively and involves the people who use pesticides, is necessary.

The second major program relates to measuring farmer's use of IPM practices. With regard to this area, the workshop stressed the need for practical, farmer-driven programs to facilitate the transition from chemical-intensive agriculture to systematically maximize the use of horticulture and biological tools to grow healthy crops and manage pests. OECD and FAO have urged Member countries to implement national pesticide risk reduction programs that use an integrated approach, encouraging participation of all important actors from a local to national level. Such programs could include instruments and activities in:

- regulations
- instruments/activities to promote safe use and improved farm management
- advice, education, and training
- monitoring and evaluation
- information exchange
- economic instruments, and
- research and development.

Member countries will work collaboratively to develop pesticide risk indicators which can be used on a voluntary basis by national governments to examine the success of risk reduction efforts in their respective countries. The priority for and implementation of any risk reduction program needs to be determined by national authorities, not by an OECD mandate.

Another project underway is the international harmonization of risk assessment methodologies for carcinogens, mutagens ad reproductive toxins. Participants in this project include the International Program on Chemical Safety (IPCS) which is an arm of the World Health Organization (WHO), European Community (EC) and the OECD countries. A recent survey of risk assessment methodologies practices by OECD and selected non-OECD countries was conducted by IPCS (Dragula and Burin, 1994)

7. Harmonization of Classification and Labeling

An Advisory Group on the Harmonization of Classification and Labeling has been working on harmonized classification and labeling systems for chemicals based on intrinsic hazard. The OECD survey results showed that all responding countries had a system for classifying plant protection products for the purpose of label warnings and precautions. In addition, all countries indicated that they classified and labeled pesticide products both for acute hazards to humans and other concerns, such as chronic human health hazards and environment effects. The focus of the work has been several health hazard categories and a single environmental proposal dealing with European Union classification "Dangerous to the Environment." It is not readily apparent from this survey about the significant differences in the classification system used by different countries for the same hazard.

North American Free Trade Agreement (NAFTA)

Harmonization efforts are moving quickly within NAFTA's Free Trade Working Group (TWG) on Pesticides. A series of bilateral and trilateral meeting were held between, U.S., Canada, and Mexico on registration harmonization. Significant progress has been made by these three government regulatory bodies. Some of the highlights are as follows:

- MRL harmonization
  - Identify and resolve MRL trade irritants with the industry and grower groups;
  - Canada and U.S. have established common geographic zones for conducting residue studies;
- Data requirements for registrations
- Study protocols for each data requirement
- Parallel Registration Review: Parallel review by U.S. and Canada of tebufenozide, an insect growth regulator;
- Registration of Heavy Duty Wood Preservatives (creosote)
- Sharing acute toxicology reviews
- Harmonizing data requirements for the registration of pheromones.

The NAFTA group has begun work to jointly review new pesticide registrations. Through its joint review, the countries will divide the work of review of a new pesticide active ingredient, which will mean less work for each participating country.

Significant short term exchange of staff and training has happened between U.S. and Canadian pesticide regulators which was responsible for building confidence and understanding, and subsequent ongoing harmonization of the registration system. At a recent NAFTA/TWG meeting, delegates pledged to accelerate the pace of work sharing. The three countries stated their aim was for work sharing to become routine by 2002. However, each country needs to achieve coordination and commitment among scientific and regulatory staff to harmonize projects.

MERCOSUR

Another regional trading block that has formed in the Western hemisphere is comprised of Argentina, Paraguay, Uruguay, and Brazil (MERCOSUR). MERCOSUR Member States are still in the process of negotiating changes in their national regulations that will result in one harmonized registration process for the group. As a first step, a positive list of active ingredients was established to allow for the free movement of products within the region. The completion of a harmonized pesticide registration system for the trading block is expected by the year 2000.

Conclusions

The above discussion clearly shows that multiple activities, albeit not all well coordinated, are happening in different regions of the world. There are still many details to be worked out before tangible evidence can be recognized. However, one thing is for sure - there has been tremendous progress on harmonization efforts in the last four years than that has happened in the last 25 years. However, at the country level, there is still much remaining to be done. Without a clear policy and commitment on harmonization, at the country level, it is possible that some countries may end up paying a high price in the months to come. The future for crop protection product registration and markets is potentially bright but real efforts to reach a consensus are needed if it is to fulfill its promise.

The harmonization of regulatory requirements is seriously under way on several bilateral, multilateral, regional, and international levels. This effort will continue since it is driven by government regulatory agencies. The challenge is to build on these successes we have had so far and achieve significant work sharing in the future.

We have come a long since the first Pesticide Forum meeting in 1992.

- Regulators and scientists in different countries have become acquainted and better understand each other's review process.

- Data review exchanges and personal exchanges are occurring fairly routinely.
- Governments are beginning to use other national or international reviews to form their hazard assessments.
- Harmonization of data review documentation/formats is occurring.
- In some cases, countries (U.S. and Canada) are jointly sharing registration burden through joint sharing of reviews.

This paper has outlined some of the activities involved in global harmonization of pesticide registration. Harmonization activities were an investment in the future. If countries invest in harmonization now, payoffs will be realized in the future. Harmonization can improve food safety through reduced residue violations and protective standards that are adhered to in all countries. It can reduce costly trade disputes. It can reduce the workload of national regulatory bodies through increased exchange of reviews. It will also improve scientific rigor through debate and the exchange of scientific information.

Harmonization could have a negative effect if the adopted standards are based on policies of pesticide use reduction, are a compilation of all existing requirements, or involve an inappropriate transfer of requirements and risk assessments. On the other hand, the industry and the grower community could benefit significantly if harmonization proceeds to the point where a core set of data requirements, study guidelines, and assessment procedures based on sound principles is widely adopted. Potential benefits could consist of a more efficient planning process, lower developmental costs, and earlier market entry with new safer technology being made available to the users in a more timely manner.
All of the data review registration harmonization efforts have common elements:

- Anticipate and prepare for international review by having
    - transparency of internal review process and
    - clear citations of data
- Public availability of data reviews is needed to facilitate exchanges, at least between national and international regulatory staff, with appropriate treatment and protection of Confidential Business Information (CBI).

If momentum is maintained, it is only a matter of time before countries begin relying on each other's reviews in lieu of conducting their own.
In the not too distant future, geographical or trade neighbors might have procedures like the European Union (EU) process whereby registration in one country can be quickly extended to other cooperating countries, with common labeling to protect health and environments and open up markets.

Literature Cited

1. Dragula, C. and Burin, G. (1994). International Harmonization for the Risk Assessment of Pesticides: Results of an IPCS Survey. Reg. Tox. Pharm., 20, 337-353
2. EC (1993). EEC Directive 91/414. Council Directive Concrning the Placing of Plant Protection Products on the Market.
3. Frawley, J.P. (1987) Codex Alimentariou - Food Safety - Pesticides. Food Drug Cosm. Law. J., 42, 168-173
4. GIFAP (1984) Pesticide Residues in Food. International Group of National Associations of Pesticide Manufacturers, Brusselles,
5. Mayburry, R.B. (1989) Codex Alimenterius Approach to Pesticide Residue Standard. J. Assoc. Off. Anal. Chem., 72, 538-541
6. Wessel, J.R. (1992) Codex Committee on Pesticide Residues - A Plan for Improved Participation by Governments. Reg. Tox. & Pharm., 116, 126-149

Chapter 17

Foreign Competition and Trade

John J. VanSickle

Food and Resource Economics Department, University of Florida, Gainesville, FL 32606-0240

Regulations controlling the use of chemicals in crop production have become increasingly important because of the effect that they may have on comparative advantage for producers competing in a global marketplace. Regulations governing chemical use are imposed to protect workers, resources, the quantity and quality of the food supply, and the environment for consumers. These regulations can influence comparative advantage across regions, changing the economic viability of producers who depend on those chemicals to maintain productivity. Methyl bromide provides a case study for impacts that regulations may have on competitiveness of producers of fresh winter vegetables. The phaseout of the use of methyl bromide by January 1, 2001, as mandated by the U.S. Clean Air Act, will have significant implications for producers of fresh vegetables during winter months. Producers impacted by these regulations may be expected to look to trade laws for protection from foreign producers who do not follow similar regulations.

International trade has become more important in recent years as an issue in policy discussions concerning pesticide regulation. The globalization of markets throughout the world makes regulations more important in determining comparative advantage for producers. Comparative advantage is the relative advantage that producers control in producing and marketing products, given the availability and opportunities for using the resources required for producing and marketing those products. Cost of production advantage is a measure of cost advantage but not a measure of comparative advantage. For example, a grower may be able to grow corn at a lower cost than growers in other producing areas, but if that grower can earn more profit producing tomatoes on that land, then he will produce tomatoes. His comparative advantage allows him to employ his resources in tomato production, leaving corn production to other growers.

Resources that are important for determining comparative advantage include natural resources, management skills and spatial location. Natural resources are those inputs provided by the environment, including land, water and climate. Labor, applied inputs (such as machinery, fertilizer and pesticides) and management skills are other inputs that can affect comparative advantage.

The environment provides the natural resources that allow producers to grow crops and products. The environment may supply inputs that are both good and bad. For example, Florida is generally blessed with ideal weather for growing and marketing fresh fruits and vegetables, but it must battle the pest pressures prevalent in warm, humid climates. Labor also provides conflicting contributions. The U.S. labor supply is much more regulated and more expensive as compared with developing countries producing similar products. But U.S. labor is generally more skilled and better equipped to be more productive than is the labor found in developing countries. These advantages offset some of the disadvantages presented by higher costs for labor.

One factor that is beyond the direct control of producers is the higher cost of inputs caused by regulations imposed by policymakers. Pesticide regulations fall within this venue. Governments impose policies to regulate inputs that have impacts on workers or the environment of their citizens. The globalization of markets has caused these policies to come under increased scrutiny. Regulations that are applied unilaterally impose higher costs on producers who have to satisfy these regulations while producers in other countries are free to produce without restrictions. There are many examples of government regulations that impact comparative advantage, including those of labor and environmental regulation. The soil and post-harvest fumigant methyl bromide provides one example of how a regulation imposed unilaterally can impact comparative advantage.

Methyl Bromide as a Case Study of Pesticide Regulation and Its Impact on Comparative Advantage

Methyl bromide is a soil and post-harvest fumigant that is critical for growing and marketing a variety of fresh fruits and vegetables throughout the world. It has had a very positive impact on the quantity and quality of these fresh fruits and vegetables. Methyl bromide is used as a soil fumigant in many parts of the world because it is a broad-spectrum pesticide with the ability to control weeds, nematodes and other soilborne pests.

The Montreal Protocol on Substances that Deplete the Ozone Layer (hereafter referred to as the Montreal Protocol) is an international agreement between member nations to oversee the manufacture and trade of ozone-depleting substances. The ninth meeting of the Parties (members of the Montreal Protocol) agreed to amend the Montreal Protocol to add methyl bromide as an ozone-depleting substance. The Montreal Protocol adopted an ozone depletion potential (ODP) for methyl bromide of 0.7 and agreed to freeze the production and use of methyl bromide at 1991 levels beginning in 1995. A total phasing out of methyl bromide use was to occur in 2005. Developing countries, as defined in Article 5.1 of the agreement, were given until 2015 to use methyl bromide with no phaseout schedule before that time.

Soil fumigation accounts for nearly 80% of the worldwide use of methyl bromide, according to figures collected by UNEP (*1*). Tomatoes and strawberries account for more than one-half of that use, with 35% and 20% of all soil fumigation applied in tomato and strawberry production, respectively. The United States accounted for more than 44% of the worldwide use of methyl bromide for soil fumigation purposes, with tomatoes, strawberries, peppers, cucurbits and eggplant accounting for nearly one-half of all methyl bromide used for soil fumigation purposes. Growers of these crops have used methyl bromide for almost 30 years and have yet to identify an alternative that is economically viable in their operations, that is, an alternative that would allow a seamless transition with improved or unchanged economic returns. If the Montreal Protocol were the only regulation imposed throughout the world, it would have an effect on the comparative advantage of those producers who rely on methyl bromide. Fresh fruit and vegetable producers in California and Florida would be especially vulnerable to those regulations for two reasons. First, the main source of competition for U.S. producers for these crops comes from Mexico, and producers in Mexico rely much less on methyl bromide than do U.S. producers. They rely on methyl bromide less because of reduced pest pressures and because of a large land base that allows them to use crop rotation pest management systems to control pests. Second, Mexico is identified as an Article 5.1 country, giving it an additional 10 years to use methyl bromide beyond the timeframe allowed for developed countries like the United States.

To compound the effect of the Montreal Protocol on U.S. producers, the U.S. Clean Air Act specifies that substances with an assigned ODP of more than 0.2 (identified as Class I ozone depletors) must be phased out of use by January 1, 2001, putting U.S. producers at an added disadvantage. The U.S. Clean Air Act was passed with the primary objective of maintaining and enhancing the quality of the earth's atmosphere. It is also the enabling legislation that requires the United States to conform to ozone regulations agreed to within the world community.

Major efforts by government and industry scientists have failed in the discovery of new chemistry with the beneficial characteristics to biologically and economically replace methyl bromide. Several studies have indicated that the schedule for phasing out the production and use of methyl bromide in the United States will have devastating consequences for U.S. producers. An early assessment of the winter fresh fruit and vegetable industry indicated that a ban on the use of methyl bromide would cause a decline in shipping point value of these crops in Florida of $620 million (*2*). The total economic impact on the state was estimated at more than $1 billion, with a loss of more than 13,000 jobs. Production of tomatoes was expected to decline 60% with strawberries expected to decline by 69%. The results also demonstrate the significant shift in comparative advantage. Because Mexican producers are afforded more time to use methyl bromide as an Article 5.1 country and because they do not rely on methyl bromide nearly as much as U.S. producers, production in Mexico that is shipped to U.S. markets is expected to expand significantly. Spreen et al. estimate that Mexican shipments of tomatoes to U.S. markets will expand by 80%, peppers by 54% and eggplant by 123% (*2*). U.S. producers lose significant market share to Mexican producers because methyl bromide regulations shift comparative advantage in favor of Mexican producers.

The 1997 meeting of the Parties in Montreal revised the schedules for phasing out the use of methyl bromide. The new schedule for developed countries calls for a 25% reduction in the use of methyl bromide by January 1, 1999; a 50% reduction in use by January 1, 2001; a 70% reduction in use by January 1, 2003; and a total ban by January 1, 2005. Article 5.1 countries have a scheduled reduction in use of 20% by January 1, 2005, and a total ban by January 1, 2015. The new schedules will have even larger implications to comparative advantage unless viable alternatives are developed.

Regulatory Solutions to Unilateral Regulations

Producers of U.S. products have expressed concerns about the impacts that regulations may have on their competitiveness in global markets. These are the concerns that are being presented as domestic producers call for the enactment of "blue and green" laws that would offset disadvantages imposed by regulations intended to protect workers and the environment. These "blue and green" laws are presented as a means of imposing higher costs on producers in foreign countries who produce products with lower standards than those in the domestic economy and then wish to sell those products in our markets. These added costs could take the form of tariffs, quotas or countervailing duties. Producers present the argument that higher standards being imposed on the production of products should be imposed on all products sold in the domestic market. Certainly, consumers concerned enough to call for regulations on treatment of labor do not discern between foreign and domestic workers. "Blue collar" laws would offset advantages afforded foreign producers because of higher costs imposed by regulations intended to protect U.S. domestic workers. "Green" laws would offset advantages afforded foreign producers because of higher costs imposed by regulations intended to protect the environment.

International trade policy has brought these issues to the table as discussions have moved forward in globalizing markets throughout the world. The Uruguay Round Trade Agreement included two important side agreements. The Agreement on Sanitary and Phytosanitary Measures (SPS) was a significant step forward in harmonizing the procedures or requirements used by governments to protect human, animal or plant life or health from risks arising from the spread of pests and diseases, or from additives or contaminants found in food, beverages or feedstuffs. The SPS Agreement gives member governments the right to maintain and enforce sanitary and phytosanitary measures as long as certain conditions are met. These include requirements that measures be based on scientific principles (Article 2.2 of the Agreement) and that they not be arbitrarily or unjustifiably imposed to discriminate against imports (Article 2.3 of the Agreement).

The second side agreement of the Uruguay Round that bears consideration is the Agreement on Technical Barriers to Trade (TBT). The Agreement on TBT allows governments to enforce higher product standards to protect the environment, health and safety of its citizens. Protection of human health or safety, animal or plant life or health, and the environment are among the objectives specifically recognized as legitimate in Article 2.2 of this Agreement, but the protection afforded in this Agreement cannot be more trade-restrictive than necessary to fulfill a legitimate objective. The Agreement requires governments to use relevant international standards as the basis for imposing domestic regulatory requirements but allows a departure from those standards for fundamental climatic or geographic factors or fundamental technological problems.

Both of these Agreements have been looked to by producers who seek relief from regulations that put them at a disadvantage to global producers who do not have to meet the same regulations. Environmentalists have also looked to these Agreements as possible avenues for regulations that restrict trade based on "product characteristics or their related processes and production methods." The SPS Agreement gives governments the ability to impose trade restrictions on products that pose risks to the human, animal or plant health or life within their territories. The TBT Agreement expands governments' rights to restricting trade on products that endanger the environment.

The globalization of the marketplace has certainly increased the sensitivity of producers who compete with differing restrictions on how they produce and market those products. Rules have been imposed through the General Agreement on Tariffs and Trade (GATT) and in domestic law to prevent unfair trade practices that give producers unfair comparative advantage. Anti-dumping and Countervailing Duty laws are means that have been provided governments to counter unfair trade practices from global competitors. These have been used to counter unfair trade practices (for example, sales below fair market value) or to counter subsidies in foreign countries that cause significant injury to domestic producers (that is, countervailing duties).

Concluding Comments

The development of new technologies and their adoption by producers will change the comparative advantage of producers worldwide. Technology is generally developed to allow producers to capitalize on resources for which they have a particular advantage in managing in their production processes. Globalization of markets and the increase in multinational firms makes the transfer of technology across borders more rapid. Because of this, regulations and financial resources available for adopting new technologies are having larger implications for comparative advantage.

The regulatory environment does not transfer across borders unless agreed to within the context of an international agreement like the Montreal Protocol. Even then, the impacts of regulations within international agreements are dependent on the importance of the regulations within countries. The Montreal Protocol and its impact on U.S. regulations will have significant impacts on U.S. producers. It should be expected that those producers will pressure policymakers to change those regulations or to finance the development of new technologies that will minimize those impacts.

References

1. United Nations Environmental Programme (UNEP). *Report of the Methyl Bromide Technical Options Committee, 1995 Assessment*; Methyl Bromide Technical Options Committee: Washington, DC, 1995.
2. Spreen, T.H.; VanSickle, J. J.; Moseley, A.E.; Deepak, M.S.; Mathers, L. *Use of Methyl Bromide and the Economic Impact of Its Proposed Ban on the Florida Fresh Fruit and Vegetable Industry*; Experiment Station Bulletin 898; University of Florida: Gainesville, Florida, 1995.

Chapter 18

Marketing Agricultural Products Internationally

Robert L. Epstein [1] and Carolyn Fillmore Wilson [2]

[1] Agricultural Marketing Service, U. S. Department of Agriculture, Mail Stop 0222, Washington, DC 20090-6496
[2] Foreign Agricultural Service, U. S. Department of Agriculture, Mail Stop 1027, Washington, DC 20250-1027

Pesticide use in food production in the U.S. is controlled by rigorous Federal and State laws. Without the availability of agricultural chemicals the quality and quantity of food to feed a rapidly growing world population would be impossible. Emerging national markets are establishing their own maximum residue limits (MRLs) following the lead of U.S., Codex or European Union standards to assure a safe and wholesome food supply. To help our Nation's farmers compete in an expanding global market, the U.S. Department of Agriculture (USDA) has relied on national statistical residue data generated by the Pesticide Data Program (PDP) to provide an overview of pesticide residues in U.S. commodities to ensure foreign markets that our food is safe. PDP was developed in 1991 as a Federal-State partnership to communicate information on pesticide residues. In 1996 PDP became an integral part of the Food Quality Protection Act, directing USDA to collect improved pesticide residue information on foods mostly consumed by infants and children to support the Government's dietary risk assessment studies.

Pesticide use in food production is controlled by rigorous Federal and State laws. Without the availability of agricultural chemicals, the quality and quantity of food to feed a rapidly growing world population would be impossible. Emerging markets are establishing their own national residue limits for pesticide residues in foods based on food consumption surveys paralleling their own dietary patterns. Where possible, emerging countries follow the lead of the U.S., Codex, or European standards to assure a safe and wholesome food supply to their respective populations. To help our farmers compete in this new environment, USDA relies on statistically-reliable residue data to provide a national overview of pesticide residues in our food. Such data are becoming an important marketing tool to expand U.S. exports into foreign markets to benefit American Agriculture. Another USDA program of considerable importance is the Sanitary and Phytosanitary (SPS) Enquiry Point at USDA, which coordinates information and comments on pesticide issues and international trade.

Rationale for Pesticide Monitoring in Foods

The Pesticide Data Program, or as it is known among the pesticide community--"PDP", had its beginnings following the "Alar in Apples" crisis in 1989. This crisis made it apparent that government needed better pesticide residue data upon which to base its decisions. USDA, with cooperation from the Environmental Protection Agency (EPA) and the Food and Drug Administration (FDA), developed a program--PDP--to provide the needed data. Initially, PDP focused on fresh fruits and vegetables, but was later expanded to accommodate processed fruits and vegetables, grain products, and fluid milk.

PDP's objectives to support issues relating to food safety are also tied to trade objectives.

- ♦ Provide the U.S. Government with high-quality data on residues in food for dietary risk assessments and address pesticide reregistration issues.

- ♦ Address the recommendations of the 1993 National Academy of Sciences' report, "Pesticides in the Diets of Infants and Children."

- ♦ Support the Foreign Agricultural Service's (FAS) international marketing of U.S. commodities.

- ♦ Provide data on pesticide residues which may affect good agricultural practices relating to integrated pest management objectives.

- ♦ Address USDA's Food Quality Protection Act responsibility.

Federal-State Cooperation. PDP is a federally funded-State operated program. There are 10 participating States: California, Colorado, Florida, Maryland, Michigan, New York, Ohio, Texas, Washington, and Wisconsin. These 10 States and States in their distribution network represent about 50 percent of the Nation's population and major regions of the country (Figure 1).

Representative Sampling. The PDP sampling system requires collection of:

- ♦ Food samples close to the point of consumption such as, terminal markets, supermarket distribution centers, and fluid milk plants, while retaining identity of product origin;

- ♦ Samples year-round based on marketplace availability using a statistically-reliable sampling protocol;

- ♦ Samples for each commodity at random without bias toward national or state origin or crop variety; and

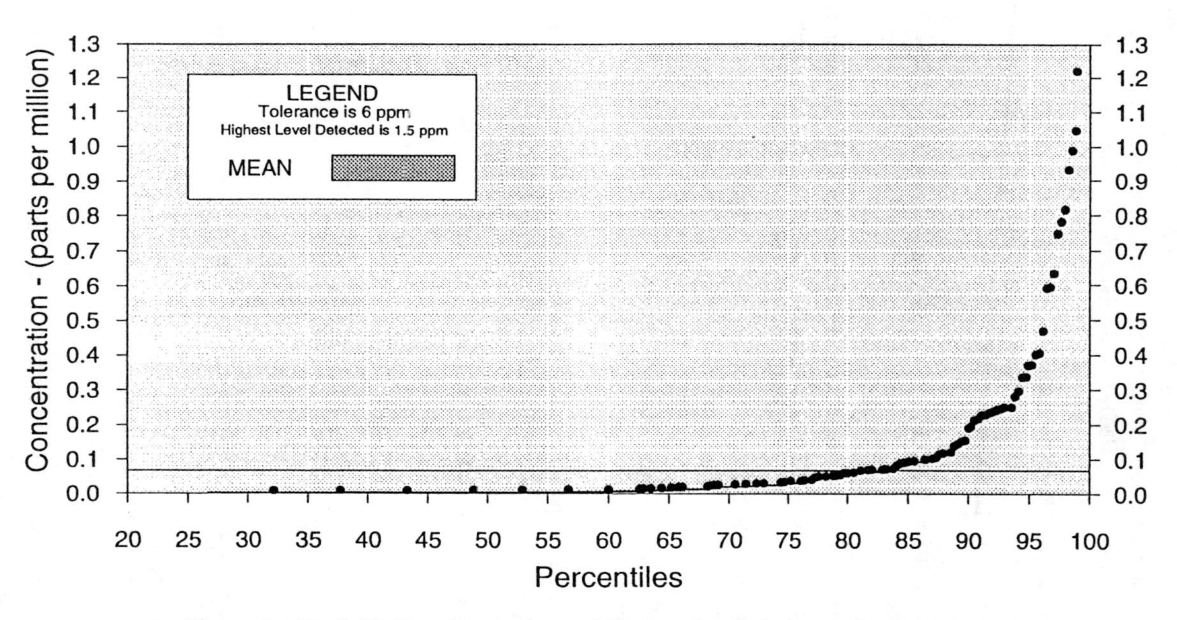

Figure 1. Pesticide Data Program participating States and state in their direct distribution network.

- ♦ Numbers of samples related to the volumes of product distributed annually by each of the 900 fresh fruit and vegetable, processed product, and fluid milk sites.

This approach to sampling provides a better picture of actual dietary exposure to pesticide residues by taking into account pesticide degradation that occurs during transit and storage. Sampling at these locations also provides information on post-harvest application of fungicides and growth regulators. Random samples are obtained in PDP based on strict protocols developed with the assistance of USDA's National Agricultural Statistics Service, enabling national residue estimates.

The sampling design resulted in a correlation between samples collected per commodity by PDP over several years versus statistical figures on imported versus domestic product availability (Table I). There is also a close correlation between PDP representative sampling and State production figures by commodity (Table II). This statistical sampling representation presents a clear and realistic picture of actual pesticide residues in U.S. commodities and plays an important role in marketing U.S. products.

Testing and Data Reduction. Collected samples are sent to State laboratories where they are analyzed for pesticides of interest to EPA. Upon arrival at the testing facility, samples are visually examined for acceptability and discarded if determined to be inedible (e.g., decayed or extensively bruised). Accepted samples are then prepared for analysis emulating practices of the average consumer. Following these practices, such as washing, removing stems, and peeling, helps to generate data which more closely represent actual human exposure to residues.

Results of residue analyses are transmitted electronically to the Science and Technology program of AMS where the data are checked, compiled, and reported for each of the more than one thousand pesticide/commodity pairs in the system. An exception to this process occurs with grain products. Grains are collected and analyzed by the Grain Inspection, Packers and Stockyards Administration of USDA--simply because it is cost effective to have this USDA agency sample products which it already has on file.

The output of PDP is the Annual Data Summary which receives national and international distribution. To date, six annual summaries were been published. Residue data are also electronically provided to EPA on a continual basis and are also available on the Internet.

Program Authority. The authorities under which PDP operates are the Agricultural Marketing Act of 1946 and, more recently, the Food Quality Protection Act (FQPA) of 1996. Title III, Section 301 © states: "The Secretary of Agriculture will ensure that the residue data collection activities conducted by the Department of Agriculture in cooperation with the Environmental Protection Agency and the

Table I. Multi-year Commodity Comparisons for Pesticide Data Program Randomly Collected Samples by Commodity versus USDA Market Data for Domestically Produced and Imported Products

		Percent			
		PDP Samples		*USDA Market Data*	
Commodity	*Year*	*Domestic*	*Import*	*Domestic*	*Import*
Fresh					
Apples	93-96	95.5	4.5	95.6	4.4
Bananas	93-95	0	100	0	100
Brocolli	93-94	97.6	2.4	96.4	3.4
Carrots	93-96	95.0	5.0	93.1	6.9
Celery	93-94	98.6	1.4	97.9	2.1
Grapefruit	93	99.8	0.2	99.0	1.0
Grapes	93-95	55.6	44.7	69.2	30.8
Lettuce	93-94	99.6	0.4	99.6	0.4
Oranges	93-96	99.0	1.0	99.3	0.7
Peaches	93-95	65.5	34.5	92.0	8.0*
Potatoes	93-95	99.0	1.0	96.0	4.0
Spinach	95-96	96.4	3.6	98.4	1.6
Canned/Frozen					
Sweet Peas	94-96	97.3	2.7	95.0	5.0
Sweet Corn	94-96	97.0	3.0	98.0	2.0

* *Includes Nectarines*

Table II. Multi-Year Comparisons of Randomly Collected Pesticde Data Program Samples versus USDA Market Data for Selected Representative Commodities in Major Agricultural States

			Percent	
State	*Commodity*	*Year*	*PDP Sampling**	*Marketing Data*
California	Broccoli	93-94	88	91
	Celery	93-94	71	84
	Carrots	93-95	68	71
	Grapes	93-96	95	96
	Oranges	93-96	78	76
	Spinach	95-96	65	63
Florida	Green Beans	93-95	37	35
	Oranges	93-96	18	20
Michigan	Apples	93-96	8	6
New York	Apples	93-96	9	8
Texas	Grapefruit	93	12	7
Washington	Apples	93-96	62	61
	Celery	93-94	4	4
	Potatoes	93-95	13	10

* *Includes Packers & Growers*

Department of Health and Human Services, provide for the improved data collection of pesticide residues, including guidelines for the use of comparable analytical and standardized reporting methods, and increased sampling of foods most likely consumed by infants and children." This statement serves as a commitment by the U.S. to provide and maintain safe and wholesome food supply, hence it is used an effective selling point, because dietary risk assessments are directed at the most vulnerable population group--infants and children.

PDP, as a result of the provisions of FQPA, is having a more significant role in providing the data needed to evaluate cumulative exposure to pesticide residues with a common toxicological effect as well as a statistically-reliable database on endocrine disruptors at minute detection levels that are needed to assess dietary risk to compromised population groups. Some known endocrine disruptors are: aldicarb, benomyl, DDT and its metabolites (byproducts), endosulfans, and parathion.

Cooperation of Market Distributors. It is important to understand that participation in PDP by companies is voluntary--not mandatory. To date, we have received excellent cooperation in obtaining the consent of terminal market vendors, managers of distribution centers, etc., to have their product sampled. Once the nature and benefits of PDP are explained to them, access to their facilities is granted for sampling purposes. AMS has provided these vendors certificates for display and letters of appreciation for their participation in PDP.

PDP operational rationale is different from regulatory or tolerance-enforcement residue programs operated by FDA and some States. PDP was neither designed to be regulatory in nature, nor to enforce established tolerances. Commodity samples are taken in commerce and products are not retained during testing. The Program was also designed to detect pesticide residues at the most minute levels possible, as technology permits, and assigns limits of detection for each pesticide and commodity pair in the testing system. This is an important premise in conducting realistic risk assessments under FQPA, because PDP facilitates consideration of non-detected residues by assigning values based on predetermined limits of detection for all pesticide/commodity combinations in the testing system. All detected pesticide residues in PDP are verified using an alternate system of analysis. It also serves as a critical factor in demonstrating that many of the pesticides dissipate in the environment and therefore should not present a food safety concern to foreign buyers.

Simply stated, PDP is not trying to find problems or residue violations, but rather, PDP is attempting to present the most statistically-reliable set of results possible. The food products and pesticides covered by PDP are selected through consultations with EPA. We try to accommodate EPA's needs to the extent permitted with existing funds. As of January 1998, 31 commodities have been considered in the PDP testing system (Table III). The testing programs include more than one hundred pesticides and residue byproducts.

Table III. A Chronological History of Commodities in the Pesticide Data Program

Start Date	*End Date*	*Commodity*	*Type*
May-91	Dec-96	Grapes	Fresh
May-91	Dec-94	Lettuce	Fresh
May-91	Dec-95	Potatoes	Fresh
Aug-91	Dec-93	Grapefruit	Fresh
Aug-91	Dec-96	Oranges	Fresh
Sep-91	Dec-96	Apples	Fresh
Sep-91	Sep-95	Bananas	Fresh
Feb-92	Mar-94	Celery	Fresh
Feb-92	Dec-95	Green Beans	Fresh
Feb-92	Sep-96	Peaches	Fresh
Oct-92	Dec-94	Broccoli	Fresh
Oct-92	Sep-96	Carrots	Fresh
Apr-94	Mar-96	Sweet Corn	Canned/Frozen
Apr-94	Jun-96	Peas	Canned/Frozen
Jan-95	Sep-97	Spinach	Fresh
Feb-95		Wheat	Grain
Jan-96		Milk	Dairy
Jan-96		Green Beans	Canned/Frozen
Jan-96		Sweet Potatoes	Fresh
Jul-96		Tomatoes	Fresh
Jul-96		Apple Juice	Processed
Sep-96			
Dec-96		Soy Beans	Grain
Dec-96	Dec-97	Peaches	Canned
Jan-97		Orange Juice	Processed
Jan-97		Pears	Fresh
Jan-97		Winter Squash	Fresh
Apr-97		Winter Squash	Frozen
Oct-97		Spinach	Canned
Jan-98		Strawberries	Fresh
Jan-98		Grape Juice	Processed
Jan-98		Corn Syrup	Processed

Data Quality and Results. A critical part of PDP's operations is the strict set of controls in place for sampling, laboratory analyses, and data reporting. Standard Operating Procedures, or "SOPs" for short, are in place for sample collections. SOPs, which are based on EPA's Good Laboratory Practices, are also in place for laboratory operations. Extensive quality control procedures are required of participating laboratories. Participating laboratories are required to determine the limits of detection and limits of quantitation for each of the commodity and pesticide pairs that they will analyze. As mentioned previously, verification is required for all initial determinations.

PDP laboratories are required to participate in a proficiency check sample program, a major recommendation by the National Academy of Sciences report. Almost monthly, prepared commodities containing pesticides of known quantities and incurred (naturally occurring) samples with pesticide residues are sent to the participating laboratories and tested under the same conditions as routine samples. The resulting data are used to determine performance equivalency among the laboratories, and to evaluate individual laboratory performance.

In addition to the rigorous quality controls, on-site visits to the States are made by our headquarters staff to ensure that sampling and laboratory SOPs are being followed. To our knowledge, no other residue testing program in the world has this degree of control over its operations, hence a very convincing argument to foreign governments on the safety of our food supply.

A question we are asked frequently is: "What are the results from PDP?" At the risk of being overly simplified, the general results from PDP can be stated as follows:

- No residue detections occur in 30 to 40 percent of the samples tested.
- When residues are detected, they tend to be at low levels--well below established tolerances.
- Presumptive violations occur primarily because pesticide residues are detected on food products for which there are no established tolerances. (This situation can arise because of crop rotation, spray drift, or possible misuse of the pesticide).
- As you might expect, results vary by specific commodity and pesticide combination.

Program Data Implications on Trade

The number of users of PDP's data has increased since the program began operations in 1991. Today, results from PDP are used by EPA, FAS, the Economic Research Service of USDA, academia, private companies, the agricultural community, environmentalists, international organizations--such as the Codex Alimentarius Commission, and global traders.

FAS forwards copies of the annual data summaries copies to all its agricultural posts in U.S. embassies and trade offices overseas. FAS has found PDP Data Summaries useful as a tool to establish a national picture of pesticide residues in foods destined for export. FAS has successfully used PDP's results to help convince foreign governments that our food is safe. As a result, foreign governments, which are major importers of U.S. commodities generally do not subject U.S. agricultural exports to the same level of consignment sampling as for some other countries nor do they make extensive demands for export certification. Korea, Indonesia, Thailand, and Taiwan are examples of this success.

Other examples of specific uses of PDP's data involve the State of California. California's Department of Food and Agriculture used PDP's data to demonstrate--particularly to Pacific Rim countries--the emphasis which our country places on the safety of our food. PDP's data were used by that department to help alleviate concerns of these countries over the pesticide residue levels in our products. As of March 1998, EPA has used PDP in 43 reregistration activities (Table IV). Many of these pesticides are classified as cholinesterase inhibitors, e.g., organophosphates and N-methylcarbamates.

To illustrate the importance of residue concentration distributions for a pesticide in a specific commodity, the percentile plot for chlorpyrifos-methyl in wheat, a major export product, is an excellent example (Figure 2). The mean residue level of 0.068 ppm is at the 80th percentile--80 percent of the samples contained chlorpyrifos methyl below 0.068 ppm and 20 percent of the samples above. The 50th percentile or median (half of the values above and half of the values below) was calculated at 0.004 ppm, the 90th percentile at 0.184 ppm, and the 95th percentile, 0.33 ppm. The limit of detection was 0.001 ppm, the lowest level administratively reportable in the PDP system. The illustration can be repeated for many of major export commodities, where a specific residue may be in question.

Customer Commitment. Those of us in USDA who are involved with PDP's operations are committed to continuously improving and strengthening the program. We have already received more information on how PDP's data have been beneficial to our users. We are working with the Department's integrated pest management (IPM) program staff to explore ways in which the program's data can be used to support IPM efforts.

Additional information on PDP can be obtained by contacting the Science and Technology program of the Agricultural Marketing Service at telephone number (202) 720-2158. Program information can also be obtained from the Internet. The PDP website is: http://www.ams.usda.gov/science/pdp/index.htm.

Table IV. Uses of Pesticide Data Program Information by the Environmental Protection Agency

43 Reregistration Activities

2,4-D	Fenthion
Abamectin	Imazalil
Acephate	Iprodione
Aldicarb	Malathion
Atrazine	Mevinphos
Azinphos Methyl	Methamidophos
Benomyl	Methidathion
Captan	Myclobutanil
Chlorothalonil	Parathion
Chlorpropham	Permethrins
Chlorpyrifos	Quintozene & Pentachlorobenzene
Coumaphos	Parathion Methyl
DCPA	Phorate
Diazinon	Phosmet
Dichlorvos (DDVP)	Phosphamidon
Dicofols	Propargite
Dimethoate / Omethoate	Terbufos
Disulfoton	Tetrachlorvinphos
Endosulfans	Thiabendazole
Ethion	Trifluralin
Fenamiphos	Vinclozolin
Fenbutatin Oxide (Hexakis)	

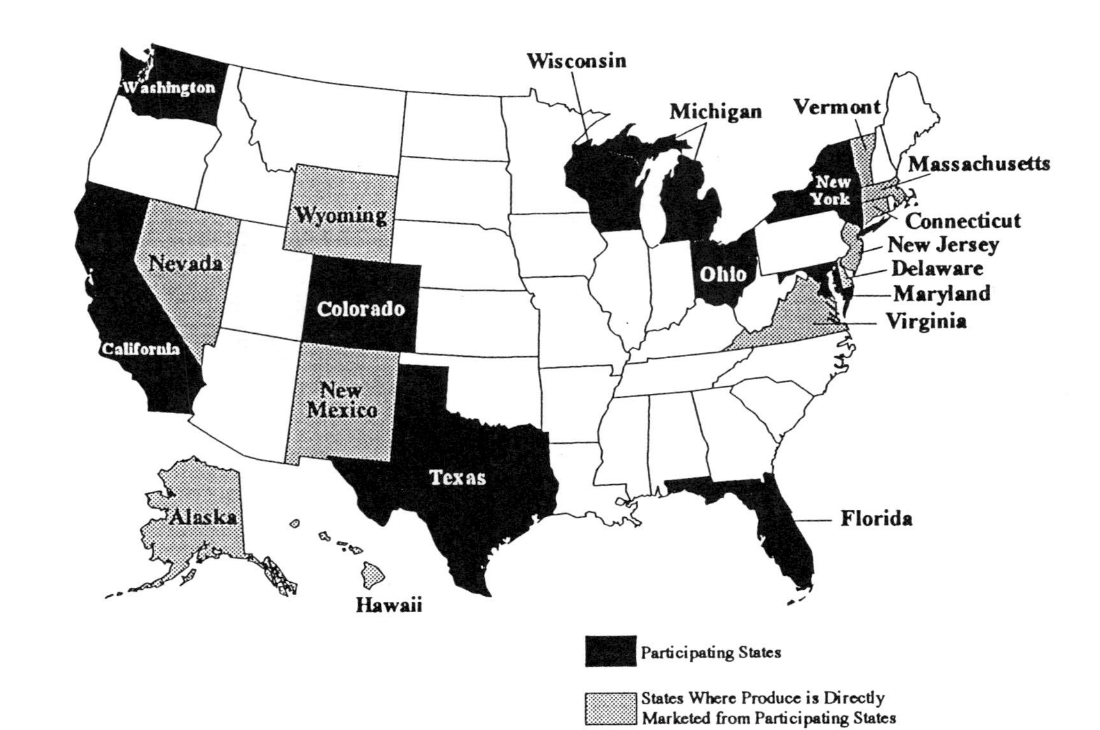

Figure 2. A cumulative percentile distribution for a representative pesticide/commodity pair (chlorpyrifos-methyl/wheat) based on 1996 crop year samples.

SPS Enquiry Point

Another aspect of enhancing U.S. competitiveness in foreign markets is the U.S. Sanitary and Phytosanitary (SPS) Enquiry Point at USDA. Presented is an overview of this FAS' program and how it relates to pesticides and international trade in agricultural products. The SPS Enquiry Point is a mandatory SPS Agreement function for the United States as a member of the World Trade Organization (WTO). From 1980 through 1994, the function was included under the old Technical Barriers to Trade (TBT) Agreement (Standards Code) of the General Agreement on Tariffs and Trade (GATT). The old TBT Agreement covered subjects that are now under either the TBT or the SPS, including pesticides.

The SPS Enquiry Point is responsible for notifying to the WTO any trade-significant U.S. proposals related to potential risk to humans, animals, plants and/or a nation's agriculture. These potential risks could arise from disease, pests, food additives, pesticide treatment, toxins, and other contaminants.

The U.S. also maintains an Enquiry Point for the WTO's Technical Barriers to Trade (TBT) Agreement at the National Institute of Standards and Technology. TBT obligations to notify do not involve risk related objectives but rather such things as labeling and quality standards.

The SPS Enquiry Point Officer reviews the Federal Register daily for trade significant proposals, confers with the proposing agency and makes a decision as to whether the proposal needs to be notified. As of mid-March 1998, the U.S. has notified 116 proposals under the SPS Agreement since the WTO came into being in January 1995. Well over one-third of the proposals, 43, were related to pesticides. Most of the notifications were proposed maximum residue limits (MRL) in foods. Some were proposed revocations or deletions because of non-support on required data.

Since the SPS Agreement only pertains to food/agriculture related products 100 percent of these foreign proposals must be reviewed and entered into the Enquiry Point data base. There are currently more than 130 member nations obligated to both the WTO SPS and TBT Agreements. These nations also notify proposed regulations, including pesticides. In the 3-years under WTO, there have been 109 proposals related to pesticides, notified by the various member nations. FAS is also responsible for reviewing all incoming foreign TBT notifications, about 30 percent of which are agriculturally-related. The rest of the TBTs are industrial.

The SPS Enquiry Point sends out an electronic newsletter every Friday evening on all agriculture-related proposals that have been notified that week through the WTO. The Enquiry Point obtains the full texts of these proposals from the other national Enquiry Points, in some cases has them translated, and provides the full text upon request.

The point of this information-sharing procedure is to offer our government agencies, trade associations, grower groups, individual exporters, and other interested parties, an opportunity to review and comment on notifications that could adversely affect trade--while the proposal is open. The proposing country has the obligation under the SPS and TBT Agreements to consider any comments provided before finalizing the regulation.

Coordination with Codex Activities. Another duty coordinated and often contributed to by the Enquiry Point Officer is comments on foreign proposals. Some of the most extensive comments over the years were on pesticide proposals, and some of our greatest successes over the years were in convincing some Pacific Rim countries to set their pesticide MRLs at Codex levels - or U.S. levels where Codex levels were not available. In the last decade, the information provided by the PDP has been especially helpful in discussions and U.S. comments around such pesticide proposals.

The SPS Agreement has wording that relies on the Codex Alimentarius Commission and other international standards-setting organizations as a benchmark. The Codex Committee on Pesticide Residues (CCPR), therefore, becomes even more important because of the additional emphasis through the SPS Agreement on these international standards. Member nations are not obligated to notify the WTO if they adopt CCPR standards exactly. Members must notify if the proposed standard does not exist under Codex or "is not substantially the same as the content of an international standard, guideline or recommendation."

It has also become increasingly important that U.S. agricultural interests be aware of and work toward having measurable pesticide standards established under the Codex system, so that agricultural products can be fairly traded internationally. The authors both serve on the U.S. Delegation to the CCPR. The authors have been very active working with the U.S. CCPR Delegation to ensure that interested parties understand the Codex system, who to contact and what is required to have a successful outcome under the system. Dr. Epstein serves as the coordinator of USDA's Pesticide Data Program. Mrs. Wilson is the U.S. SPS Enquiry Point Officer. For further information on the SPS Enquiry Point, the newsletter, or the CCPR information letter please contact Mrs. Wilson at: Phone: (202) 720-2239; Fax: (202) 690-0677; or E-Mail: wilsonc@fas.usda.gov

Chapter 19

Current Status of Domestic and International Controls for Methyl Bromide and the Status of Alternatives

Ralph T. Ross

Plant Protection and Quarantine, Animal and Plant Health Inspection Service, U. S. Department of Agriculture, Washington, DC 20250

Methyl bromide (MB) is a versatile highly effective, fast acting fumigant employed in a number of ways to kill organisms destructive to plants. Its use is important for the movement of commodities in international trade for the disinfestation of pests. The compound is unique in that it provides a wide range of pest control, may be applied to a broad spectrum of both food and non-food commodities, can be used for fumigation of large and small quantities of materials, and, when applied properly, leaves no residues of a toxicological significance. Recently, this compound has come under scientific scrutiny and has been identified as a potentially potent ozone-depleting chemical. As a result, countries operating under the Montreal protocol (MP), an international treaty for the international control of ozone-depleting substances (ODSs) and under the auspices of the United Nations Environment Program (UNEP), will be restricting its use. Many countries will be eliminating it altogether. Its limited use and/or potential phase-out will have severe econominc implications on agricultural production and trade unless alternatives become available which are efficacious and economical.

Since 1991, methyl bromide (MB) has come under scientific scrutiny and has been identified as a potentially potent ozone-depleting substance (ODS). As a result, countries operating under the Montreal Protocol (MP) will be restricting its use or eliminating its use altogether. For example, the United States Environmental Protection Agency (USEPA) has published a final rule that will terminate total production and consumption in the U.S. by 1 January 2001*(1)*, whereas developed countries operating under the MP will be eliminating production and consumption in the year 2005 *(2)*. Imposing limitations on its use and/or a total phaseout will have severe economic effects if viable alternative treatments are not available.

In addition to its worldwide use, MB is one of the few fumigants left for insect disinfestation. It is the only remaining fumigant for commodity treatment for quarantine. Commodity treatments represent approximately ten percent of its total use in agriculture *(3)*. No other treatments are available which would provide the same physical and chemical characteristics as MB and that would make them as useful as broad scale alternative commodity treatments, including quarantine treatments: that is fast action (fumigation times of 2-24 hours depending on commodity); ease and flexibility of application; and gaseous/efficacious at a broad range of temperatures.

The Montreal Protocol

In 1985 the Vienna Convention under the auspices of the United Nations Environment Program (UNEP), laid the framework for addressing substances which deplete the ozone. In response to the growing evidence that chlorine and bromine could destroy stratospheric ozone on a global basis, the international community in 1986 negotiated the MP *(4)*. The MP limits the production and consumption of specific sets of ODSs. Significant scientific advances have continued, and reports indicate a more rapid rate of ozone depletion than previously believed and that "anthropogenic sources of MB are significant contributions to stratospheric ozone-depletion *(5,6)*."

At the Fourth Meeting of the Parties to the Montreal Protocol, in Copenhagen, Denmark, 23-25 November 1992, additional adjustments were made, including an amendment for MB for developed countries. The amendment proposed: (a) to add MB to the list of controlled substances with an assigned ozone depleting potential (ODP) of 0.7; (b) to freeze production and consumption in 1995 at the 1991 levels; (c) to exempt quarantine and preshipment uses from the 1991 freeze in production and consumption; (d) to conduct a two years in-depth study on MB uses and alternatives; and (e) to re-evaluate the science in 1995 and the in-depth study *(7)*.

At the Seventh Meeting of the Parties, in Vienna, Austria, 5-8 December 1995, actions were taken to strengthen the overall controls for MB. These actions included: (a) a change in the listed ODP from 0.7 to 0.6; (b) a developed country phaseout for production and consumption on 1 January 2010 which will be preceded by two interim reductions, 25 percent 1 January 2001 and an additional 25 percent on 1 January 2005; and (c) a developing country freeze on production and consumption commencing 1 January 2002 at the average of the 1995-1998 production levels *(8)*. These actions did not affect the 1992 Copenhagen exemptions for quarantine and preshipment applications *(2)*.

The Ninth Meeting of the Parties further strengthened MB controls *(9)*. For developed countries the phaseout date was moved up from 2010 to 2005 with interin cuts of 25 percent in 1999; 25 percent in 2001; and 20 percent in 2003. For developing countries an agreement was reached for a phaseout date in 2015 which allows a ten years grace period after a complete phaseout has taken place in developed countries. Exemptions for quarantine and preshipment uses remain unchanged for both developed and developing countries. Two additional exemptions were approved by the Ninth Meeting of the Parties. These were critical and emergency uses which will not go into effect until after a total MB phaseout. The exact definitions and criteria for identifying critical and emergency uses are currently under study by UNEP's Technology and Economic Assessment Panel (TEAP). However, in general terms, a Party may apply for a critical use provided that it can show that there are no technically and economically feasible alternatives or substitutes available to the user, and without the approved critical use, will result in significant market disruptions. Further production of MB will be permitted only if: (a) no technically and economically feasible alternatives are available; (b) no MB is available from existing stocks; and, (c) the Party must demonstrate that appropriate effort is being made to identify alternatives *(9)*.

An emergency MB use is "...to allow a Party, upon notification of the Secretariat, to use, in an emergency situation, consumption of quantities not exceeding 20 tons of MB for critical uses. The Secretariat, in consultation with the TEAP, will evaluate the use according to the 'critical MB use criteria' and present this information to the next meeting of the Parties for review and appropriate action by the Parties *(9)*."

U.S. Clean Air Act

Section 602(e) of the USCAA states: "Where the ozone-depletion potential of a substance is specified in the Montreal Protocol, the ozone-depletion potential specified for that substance under this section shall be consistent with the Montreal Protocol" *(10)*. Therefore, the action taken under the Montreal Protocol provided the legal basis for USEPA to publish rulemaking for MB in the U.S. Federal Register in December, 1993 *(1)*. These actions included: (a) listing MB with an ODP of 0.7; (b) freezing production and consumption on 1 January 1994 at the 1991 levels; (c) classifying MB as a Class I ozone depleting chemical; (d) terminating production and consumption on 1 January 2001; and not requiring MB treated products to be labeled.

Comparison of Actions for MB under the USCAA and the MP

Regulatory provisions for MB under the USCAA are more stringent than those contained in the MP. The primary objective of the U.S. at the Ninth Meeting of the Parties was to persuade other Parties for a global MB phaseout in 2001 or for Parties to take actions for MB that would be consistent with or as close to as possible the USCAA; this would have provided a level or nearly level playing field for all Parties. However, most countries were not prepared to go that far, particularly developing countries. Their decision was based primarily on the importance of MB to their respective country's economies and the lack of available alternatives. After considerable debate, developed countries agreed to a MB phaseout on 1 January 2005 with two 25 percent interim reductions in 1999, 2001, and another 20 percent reduction in 2003, a total of 70 percent.

In spite of the additional actions taken by the Protocol to strengthen MB controls, there remains a regulatory gap between the Protocol and the USCAA. The Protocol exempts quarantine and preshipment uses; the USCAA authorizes no MB exemptions. Another more obvious difference is the distinction between Class I and Class II ozone depleting substances and the mandatory phaseout dates required under the USCAA. Class I and Class II ODSs are based on the numerical number of their respective ODPs, and the threshold number which separates the two classes is 0.2, (i.e., chemicals with ODPs greater the 0.2 are Class I; less than 0.2, Class II. Class I ODSs must be phased-out seven years subsequent to the listing date); Class II, by the year 2030. The Montreal Protocol does not list ODSs by classes and there are no mandatory phase-out dates; phase-out dates are determined by consensus vote by the Parties.

Impact on U.S. Agriculture

Because of the important uses of MB and their roles in agricultural production and trade, the phaseout of MB in 2001 under Title VI of the USCAA is of vital significance. This U.S. law is more restrictive than the provisions for MB under the MP that governs the rest of the world. In particular the MP allows longer phaseout schedules and provisions for essential uses and exemptions. The differences between the domestic and the international regulations has caused a profound concern among agricultural producers, processors, and those engaged in international trade. U.S. farmers are concerned that, if adequate alternatives are not available, they will be put at a significant competitive disadvantage in international agricultural trade when the U.S. phaseout takes effect.

MB is particularly important for quarantine treatments because of its effectiveness against a large variety of indigenous and non-indigenous pests and because it can be easily and economically applied to both small and large shipments or storage. U.S. regulations require that a wide array of imported food and non-food commodities be fumigated with MB as a condition of entry. In addition, a number of commodities exported by the U.S. must be fumigated with MB in order to comply with quarantine requirements of recipient countries. A critical quarantine use of MB to U.S. agriculture is its role as the only practical emergency treatment to move commodities out of areas quarantined for outbreaks of exotic pest insects such as the Mediterranean fruit fly.

The largest use for MB is as a soil fumigant and for intensive production of high value crops such as strawberries, tomatoes, cucumbers, peppers, melons, and eggplant. The 1993/94 production values for these six commodities using methyl bromide for preplant treatment were $2.4 billion.*(11)*. In addition to these six commodities, a 1993 USDA assessment report showed an additional 15 commodities for which methyl bromide was also important in their production *(13)*.

Stored agricultural food products include a wide variety of dry foodstuffs, principally cereal, grains, oilseeds and legumes, grain products, dried fruit and nuts and other durable products such as, timber and timber-containing products, and various artifacts. These products are often stored for long periods of time and are treated with MB for control of a number of domestic pests. Insect and mite pests can breed on these materials during storage. Pests may also be present at time of harvest, and persist in storage or during transportation. Control of pests infesting stored commodities is essential to keep commodity losses to a minimum, to maintain quality and prevent damage, and prevent the spread of pests between countries. In 1993/94 the estimated value of dried fruits and nuts alone was in excess of $4 billion.

Structural fumigation of food production and storage facilities (mills, food processing, distribution warehouses), non-food facilities (dwellings, museums), and transport vehicles (trucks, ships, aircraft, rail cars) are very reliant on MB for control of a large number of pests. It is used either on an entire structure or a significant portion of a structure. Fumigation is utilized whenever the infestation is so widespread that localized treatments may result in re-infestation or when the infestation is within the walls or other inaccessible areas.

Agricultural exports consistently make a large positive contribution to the U.S. balance of trade. The USDA Economic Research Service's statistics for the fiscal year 1993/94 showed the value of U.S. exports to the world market for apples, cherries, peaches/nectarines, and strawberries was $650 million; cotton, $2.3 billion; oak logs, $130 million; and walnuts (in shell) $86 million. The export market values for these commodities to countries requiring MB treatment totaled $282.8 million ($101, $106, $24 and $1.8 million respectively).*(11)*

The current extent and importance of methyl bromide use and the potential impacts that the 2001 phaseout poses for American agriculture necessitate a major effort to ensure that American farmers can continue to raise and market their crops. USDA has directed its resources and expertise, with the support of Congress and in cooperation with growers, to conduct an ambitious research program to identify and develop alternatives to control the pests currently controlled by MB.

Summary

USDA has placed a high priority on dealing with agricultural concerns while contributing to the protection of the global environment There are three areas where USDA is working to develop solutions that meet both of those needs in dealing with the MB issue. These are discussed in the following: *(13,14)*.

Research. USDA's Agricultural Research Service (ARS) has for many years devoted significant research resources to approaches with potential for replacing MB. Since the USEPA announced the phaseout for MB, ARS increased its efforts to find alternatives. ARS is seeking alternatives for MB through research on a variety of approaches which include; 1) new cultural practices, 2) improved host-plant resistance to pests and diseases, 3) biological control systems using beneficial microorganisms, and 4) less harmful fumigants. For postharvest treatment, alternatives being investigated include: a) creation of pest-free agricultural zones, b) physical methods such as hot or cold treatment or storage in modified atmospheres, c) alterative fumigants d) MB trapping and recycling technologies, e)biological control, and f) systems approach.

Spending for ARS research on methyl bromide alternatives increased from $7.4

million in FY 1993 to the $13.9 million included in the current appropriation for FY 1996. This spending supports 42 scientists years involving 46 projects. This research is augmented by research from grower groups and EPA.

USDA recognizes that there are a number of real world factors that affect our ability to find alternatives. We recognize that alternatives for a wide variety of crop applications spread over a diverse set of geographic conditions will have to be found and that no single practice will substitute for all those uses. We also know that a genuine alternative for farmers must be efficacious, cost effective, logistically possible, and available for efficient incorporation into standard agricultural practices. In addition, the approval process for a new use or new product takes time for registrants to conduct and submit the required studies that EPA must review by the latest standards. Finally, securing approval of quarantine practices by importing countries has typically taken years of negotiation.

The Montreal Protocol. USDA has actively participated in the development of United States Government positions for international deliberations by the Parties to the Montreal Protocol. USDA worked within the delegation to help level the playing field for U.S. producers in the international arena, by pressing for a global phaseout at the Vienna meeting that would require that all countries meet the same standards. Although the Parties did not ultimately support that position, important progress was made. Developed countries agreed to a phaseout schedule and a freeze on developing country use was adopted. While these measures fell short of the U.S. position, they do represent a universal commitment to international controls and the first steps toward a worldwide phaseout.

Administrative solution. Despite the progress made internationally there remains a disparity between the USCAA controls on MB in the U.S. and the controls affecting the rest of the world under the MP. As a result, USDA is very concerned that, if adequate alternatives are not available, U.S. farmers will be put at a significant competitive disadvantage in international agriculture and trade when the U.S. phaseout takes effect in 2001. The Clinton Administration has indicated its willingness to work with Congress and other stakeholders to craft a reasonable solution limited to resolving the concerns for the competitiveness of U.S. agriculture and trade by assuring the continued availability of MB where it is needed because of the lack of acceptable alternatives.

If we come to a successful and responsible solution, some important principles must be incorporated into any legislation. First of all, it must protect American agriculture and trade from being put at a competitive disadvantage. Second, it must provide sound protection of the global environment. Third, it must retain the incentives for research on alternatives. Fourth, it must not result in a cumbersome or unworkable administrative process. Finally, it must not undercut international agreements.

Literature Cited

1. United States Federal Register, 40 CFR Part 82, December 10. 1993.

2. UNEP, *Report of the Seventh Meeting of the Parties to the Montreal Protocol on Substances that Deplete the Ozone*, UNEP/OzL.Pro., December 1995.

3. Methyl Bromide Science Workshop Proceedings, Washington, D.C. (June 1992). References cited therein.

4. UNEP, *Handbook for the Montreal Protocol on Substances that Deplete the Ozone Layer: 1996.* United Nations Environment Program, Fourth Edition, 1996.

5. WMO, *Scientific Assessment of Stratospheric Ozone: 1991,* World Meteorological Organization Report No. 25, 1992.

6. WMO, *Scientific Assessment of Stratospheric Ozone: 1994,* World Meteorological Organization Report No. 37, 1995.

7. UNEP, *Report of the Fourth Meeting of the Parties to the Montreal Protocol on Substances that Deplete the Ozone*: UNEP/Ozl.Pro., November 1992.

8. UNEP, *Report of the Seventh Meeting of the Parties to the Montreal Protocol on Substances that Deplete the Ozone*, UNEP/Ozl.Pro., December, 1995.

9. UNEP, *Report of the Ninth Meeting of the Parties tot he Montreal Protocol on Substances that Depelet the Ozone,* UNEP/Ozl.Pro., September, 1997.

10. United States Clean Air Act, Public Law 159, Title VI-Stratospheric Ozone Protection.

11. United States Department of Agriculture, Economic Research Service, Statistical Data, 1993/1994.

12. USDA, Economic Research Service, *Economic Effects of Banning Methyl Bromide for Soil Fumigation*, Report No. 677, March 1994.

13. Elworth, Larry , *U.S. Department of Agriculture, Testimony Before the Subcommittee on Health and Environment, Committee on Commerce, United States House of Representatives,* August 1, 1995.

14. Elworth, Larry , *U.S. Department of Agriculture, Testimony Before the Subcommittee on Health and Environment, Committe on Commerce, United States House of Representatives,* January 25, 1996.

Chapter 20

Pesticide Disposal in Developing Countries: International Training Course to Manage Risks from Obsolete Pesticides

Janice King Jensen [1], Kevin Costello [1], and Kay Rudolph [2]

[1] **Office of Pesticide Programs, U. S. Environmental Protection Agency, 401 M Street S. W. (7507C), Washington, DC 20460**
[2] **Pesticides and Toxics Program (CMD-4), Environmental Protection Agency, Region 9, 75 Hawthorne Street, San Francisco, CA 94105-3901**

Many developing countries have accumulated large stocks of unwanted pesticides that require disposal. These stocks are often improperly stored near environmentally-sensitive areas in deteriorating, leaking containers. In cooperation with the Food and Agriculture Organization of the United Nations, the U.S. Environmental Protection Agency is developing the international regional training course "Pesticide Disposal in Developing Countries." This course provides a technical and legal framework for decision-makers and addresses ways to prevent the future accumulation of unwanted stocks. This paper describes the first two deliveries of the course (Honduras in May 1997 and Indonesia in December 1997), discusses improvements to the course, and evaluates the course as a tool for managing global risks associated with unwanted pesticides.

Many countries have large quantities of unwanted pesticide stocks that have been accumulated over the last thirty years. The Food and Agriculture Organization of the United Nations (FAO) estimates that there are about 15,000 tons of obsolete stocks in Africa that require disposal (*1*). Based on rough estimates from Latin America, the obsolete stock problem appears to be similar in scale to African countries, in the range of 100 tons to 1,000 tons per country. In Poland, the problem is significantly larger. An estimated 60,000 tons from former state-run farms now require disposal, and of that, an estimated 10,000 tons are stored in underground tombs near drinking water reservoirs (*2*). The problem in Ukraine is smaller but typical of countries in the former Soviet Union — an estimated 22,000 tons of canceled, expired, or unlabeled pesticides are located at more than 4,000 sites across the country (*3*).

Because of the scale of this problem, outside help is usually required to dispose of unwanted stocks of pesticides. Developing countries rarely have the expertise, money or disposal facilities to solve the problem themselves. The risks of inaction are high, and increase with time. Many of the unwanted pesticides described above are in deteriorating containers, which increases the risk of air, soil or water contamination and makes reformulating or relabeling the stocks for another use more difficult.

FAO Disposal Program

The FAO has pioneered the disposal of large stocks in developing countries. Since 1994, FAO has operated a project to dispose of obsolete pesticide stocks in Africa and the Near East. Under the project, FAO collects and compiles data on obsolete stocks in those regions, produces technical guidelines, conducts pilot disposal operations, and facilitates and coordinates international efforts to launch disposal operations. Through this effort, FAO has completed three pilot disposal projects in Yemen, Zambia, and the Seychelles. FAO estimates that it costs between $ 3,000 - $4,500 per ton to export obsolete pesticides to a developed country for disposal in a dedicated high temperature incinerator (*4*)

Other agencies have become involved in solving the obsolete stock problem in Africa. GTZ (the German Agency for Technical Cooperation), the Government of the Netherlands, Shell Chemical Company, and the US Agency for International Development (USAID) have completed pilot disposal operations in Niger, Zanzibar, Mozambique, Madagascar, Tanzania, and Mauritania. More comprehensive disposal operations are underway in Gambia, Madagascar, Mali, Mozambique, Senegal, and Tanzania. Disposal operations are in the early planning stage in Botswana, Eritrea, and Ethiopia, funded by a coalition including GTZ, FAO, Denmark, the European Union, USAID/Rhone-Poulenc, the Netherlands, Denmark, and the agro-chemical industry, led by the Global Crop Protection Federation (*5*).

The European Union is also becoming involved. The EU provides aid to 69 African, Caribbean, and Pacific countries using an agreement called the Lomé Convention. In 1993, these Lomé Convention countries expressed concern about the obsolete stocks problem and passed a resolution that called for support in disposing of obsolete stocks in their countries. It also called on the EU to take necessary measures to avoid the further accumulation of such stocks.

FAO provides pesticide disposal training to developing countries on a case-by-case basis, usually in advance of a disposal operation. In the hope of expanding their ability to disseminate this important information beyond their limited resources and African mandate, the FAO has produced the following series of three provisional technical guidelines:

- *Disposal of bulk quantities of obsolete pesticides in developing countries, 1996*; (*6*)
- *Pesticide storage and stock control manual, 1996*; (*7*) and
- *Prevention of the accumulation of obsolete pesticide stocks, 1995 (8)*.

The documents draw on FAO's experiences in Africa and offer valuable advice on issues to consider and common errors to avoid. They are the basic reference works for any developing country undertaking a disposal project.

International Training Course on Pesticide Disposal

Based on the global need for additional training tools that can be used to assist countries and regions make better decisions to reduce the risks associated with obsolete pesticides, the US Environmental Protection Agency developed an international training course titled "Pesticide Disposal in Developing Countries." Designed in collaboration with FAO, the EPA course is based on the three Guidelines and builds on the significant progress and lessons learned to date.

"Pesticide Disposal in Developing Countries" is one of 22 courses EPA offers internationally (*9*). Previous courses have ranged in content from general subjects such as "Principles of Environmental Policy" and "Principles of Pollution Prevention," to more specific subjects, such as "Medical Waste Management." Many of these were developed in the early 1990's in response to the demand throughout Central and Eastern Europe for environmental training, technical assistance, and information. The courses are now also being successfully presented in other regions of the world.

The "Pesticide Disposal in Developing Countries" course is delivered by EPA employees, using a facilitated, train-the-trainer approach. The course is designed to be delivered twice in a particular region. Two or more attendees chosen from the first delivery can then assist in facilitating the second delivery. After the second delivery, EPA expects to "hand-off" the course to local facilitators who would then provide future deliveries of the course. This method is intended to facilitate the development of national and regional expertise.

EPA recommends that a mix of participants attend each workshop. Ideally, six to eight countries from a region would participate, with three or four attendees from each country. Suggested participants from each country would include a senior-level decision-maker from the pesticide regulatory agency, the technical expert tasked with disposing of unwanted pesticides, and one or two representatives from industry, the media, a university, or an environmental organization.

The four day course introduces participants to the various options available to them for the disposal of bulk quantities of pesticides, and provides a basic technical, legal, and logistical framework for making decisions. The course provides this information both through classroom lectures and practical exercises. Lecture materials teach participants how to manage risk in several ways: (1) by defining risks by inventorying stocks to identify products and quantities, (2) by identifying the risks of not taking action, (3) by mitigating risks to human health and the environment during actual disposal operations, and (4) by avoiding risk by preventing the build-up of future obsolete stocks. The practical exercises include a hypothetical case study and a visit to a local pesticide storage facility.

Classroom Lectures

Risks from Inaction. The workshop begins with an overview of the risks that obsolete stocks of pesticides pose to human health and the environment, to raise awareness of the hazards of inaction. Often, the approach developing countries have taken with obsolete pesticides has been to ignore them. The disposal of bulk quantities of pesticides is difficult and expensive, and few developing countries have the infrastructure to safely dispose of them. Developing countries also often have competing priorities for scarce resources, such as for health and education programs.

Stored stocks pose several risks. Contamination can occur through leakage of the pesticides onto the ground or into runoff when it rains. People and animals can be poisoned through direct contact with the pesticides, inhalation of the vapors, drinking contaminated water, or eating contaminated foods. Some pesticides (especially certain fungicides) become unstable when improperly stored and may spontaneously combust.

Typical hazards associated with pesticide storage sites include:

- noxious fumes
- illness or death to humans, and domesticated fish, birds, mammals
- illness or death to wildlife, migratory species
- food chain contamination
- habitat destruction and threats to biodiversity
- contaminated groundwater, drinking water, irrigation water

Exposure to pesticides can occur as a result of leaking containers, spillage, fires, burial, floods, inappropriate use, accidents in transport, drift, runoff, volatilization, and improper disposal (e.g., burial).

The EPA module illustrates the above points through slides and examples drawn from past disposal projects: in Zambia, where obsolete stocks stored in leaking containers above an aquifer contaminated the water supply of the capital, Lusaka; in Yemen, where 30 tons of endrin and dieldrin buried in a 1980's irrigation project have dispersed into soil and water, contaminating an estimated 130 tons of material, and threatening to contaminate the Red Sea through seasonal floods; and in Nepal, where organophosphates were buried in their original containers in an area where erosion in the rainy season now threatens contamination of surface waters (*10*).

In addition to acute risks such as leaking containers, spills, fires, and floods, which most participants readily recognize, the course also stresses regional and global risks from obsolete stocks. Improper storage or disposal of pesticide stocks can lead to contaminant transport across national boundaries via ground-water or surface-water supplies. Atmospheric transport of persistent pesticides like DDT and dieldrin can spread contamination over even greater distances. These concepts have been more difficult to teach, because the effects of these types of contamination are not as dramatic as those of localized, acute events. However, prevention of long-range transport of pollutants is a topic of current international concern. The United Nations Economic Commission for Europe, for instance, recently completed negotiations on a legally-binding regional protocol on persistent organic pollutants (POPs) under the Long-Range Transboundary Air Pollution (LRTAP) convention. We are continuing to refine the materials for this segment of the training to share this information more effectively with developing countries.

In addition to describing the potential risks of storing obsolete stocks, lectures stress the importance of identifying site-specific problems, to determine, for example, what steps would need to be taken to prepare the site for disposal (e.g., roads to provide access for equipment), and whether the site should be stabilized to protect against immediate hazards to health and the environment. The course reviews the FAO-recommended methodology for inventorying stocks and identifying site-specific hazards — information which is critical to prioritizing projects for disposal.

Controlling Risks During Disposal Operations. A prime goal of the workshop is to emphasize the importance of controlling risks during a disposal operation, particularly the risks to workers involved in the operation. The course examines the risks of each facet of the disposal operation, from conducting and evaluating inventories, through selecting management disposal options, disposing of empty containers, working in storage sites, and stabilizing and cleaning up storage sites. The three most significant issues covered are:

- site preparation to control risks (e.g., the importance of ventilating the storage building before allowing entry);
- protective clothing (including examples of different types of respirators and gloves, the benefits and hazards of different materials, and the dangers of using old or contaminated equipment); and
- worker training (especially critical where circumstances limit the effectiveness of other protections).

Lecture materials also stress precautions that must be taken when transporting hazardous materials such as obsolete pesticides across national boundaries. Participants are informed of the rights and responsibilities of developing countries under legal instruments such as the Basel Convention and the Prior Informed Consent (PIC) guidelines proposed by the FAO and United Nations Environment Program. Countries that are party to these agreements have agreed that pesticide transboundary disposal operations will be conducted only between fully informed parties that are committed to the proper disposal of hazardous materials. In addition, lecture materials summarize United Nations labeling requirements for the overseas shipment of hazardous materials. Although there is limited time to introduce these international conventions to course participants, EPA hopes to stress that these precautions are crucial to avoiding the global risk posed by the improper transport of hazardous materials across national boundaries.

The course also discusses how to develop a communication strategy to inform the public about the operation — both to address the public's concerns and to provide nearby residents necessary information on precautions to follow while the disposal operation is underway. Disposal operations that otherwise have been carefully well-planned can be delayed or postponed by workers or nearby residents who have not been well-informed.

Prevention -- Key to Long-term Success. Disposing of existing stocks of obsolete pesticides solves only half the problem — preventing the accumulation of future stocks is critical to long-term success. The course reviews sources of accumulation, and initiates discussion about the roles of government agencies, aid agencies, and industry in preventing future stocks.

While pesticides have played an important role in controlling pests for increased food production, there are many factors that have led to the buildup of unwanted stocks. The most common of these include:

- excessive donations of pesticides by aid agencies or development banks
- purchase or donations of unsuitable products or impractical package size
- prohibition of use due to regulatory action or policy decision
- overstocking of products with a short shelf-life

- poor packaging, with missing or incomplete labels
- removal of subsidies (i.e, increases in price cause a drop in demand)
- lack of in-country facilities for laboratory analysis
- lower than expected pest incidence

Practical Exercises

The course supplements the information provided through the lectures and reading materials with practical exercises that give attendees the opportunity to apply what they learn. The first of these is a case study that challenges attendees to develop a pesticide disposal plan for a mythical country that has significant stocks of obsolete pesticides, but few resources with which to dispose of them. The case study requires participants to conduct and evaluate pesticide inventories; select management and disposal options for the pesticides, their containers, and any contaminated materials; consider the safety of workers entering storage sites; and develop a communication strategy. Participants are grouped into teams which compete to propose the best solution to the "Minister of Agriculture" on the third day of the class.

This role play-exercise is beneficial in several ways. First, it requires the participants to work through the various steps of a pesticide disposal operation in sequence, and to give some thought to some of the problems and limitations that might be encountered during such an operation. Second, it familiarizes participants with reference materials from the FAO and other organizations that are invaluable when planning a disposal operation. Finally, by working in teams with colleagues from other ministries or countries in the region, each participant can make contacts that might prove useful for planning actual disposal operations, or for other regional endeavors.

The course will ideally also include a field trip to an actual obsolete pesticide storage site in the vicinity of the lecture facilities. If official permission can be obtained, course attendees will use this opportunity to get hands-on experience in preparing a rudimentary inventory of obsolete stocks. Such a site visit allows the attendees to personally witness the effects of prolonged storage, and to better understand the logistical difficulties of rehabilitating a warehouse of failing pesticide stocks.

Honduras — May, 1997

The first international delivery of the course was held at the Panamerican Agricultural School (PAS) in El Zamorano, Honduras, 36 km east of the capital Tegucigalpa. The course was co-sponsored by the (CCAD), under a technical assistance agreement with the EPA.

Forty-two attendees from seven Central American countries (Honduras, Costa Rica, Nicaragua, Panama, Belize, Guatemala and El Salvador) participated in the course. The majority of participants were civil servants representing Ministries of Health or Agriculture, which have jurisdiction over various aspects of pesticide registration and control in their respective countries. Along with several representatives from the agrochemical industry and academia, the group possessed extensive knowledge on the use and registration of pesticides in Central America. Since this was the first delivery of the course (other than a dry run at EPA

Headquarters for EPA facilitators) the participants were asked for extensive comments on course materials and how materials were presented.

What Worked. The coverage of the Central American region represented by the participants was ideal. The mix of countries and agencies resulted in a truly regional picture of the problem obsolete stocks might pose to Central America, and formed the basis for a regional network of contacts to draw on for future disposal efforts. Several participants were identified as excellent candidates to be facilitators at the second delivery of the course, which is planned to take place in the fall of 1998 at OIRSA headquarters in El Salvador.

The PAS provided an excellent atmosphere in which to deliver this course. The W.K. Kellogg Center, where both the classrooms and lodging were located, offered few distractions. The participants responded by regularly working after hours on course exercises.

The most successful and popular part of the course was the pesticide disposal case study, set in the mythical country of Paraiso. The teams of participants worked through the evenings on the case study in a spirit of friendly competition, and their presentations to the "Paraiso Minister of Agriculture" on the third day of the course were well thought-out and surprisingly polished, given the time available to prepare them.

Marco Gonzales, Legal Advisor to CCAD, delivered a lecture on regional legal requirements concerning the disposal of obsolete pesticides. By setting the EPA lectures on international legal considerations for pesticide disposal against a local context, participants were better able to gauge the relevance of these international issues to Central America.

The participants had a practical opportunity to take a pesticide inventory on the fourth day of the course, when we visited a government storeroom of aging pesticide stocks in Tegucigalpa. The manager in charge of the stocks explained that the pesticides stored in the warehouse, some of which are banned internationally, have been at the location for as long as 20 years. He described the warehouse of corroded and broken pesticide containers as a "time bomb" located in the nation's capital. After visiting this site, the class toured a newer, model storage facility recently built outside the capital with funding from international aid organizations (*11*).

What Didn't Work. While the participants were at the appropriate professional level for potential facilitators for future deliveries of the course, the higher level decision-makers who will manage disposal efforts were not present at the workshop. The impact of the course will hinge in part on how effectively the message about pesticide disposal is relayed to these decision-makers.

Translation of English-language presentations impeded discussion at times. The school had provided working equipment for simultaneous translation between Spanish and English. However, bilingual interpreters experienced enough to keep up with the presentations and participants' questions were not available.

The session in which participants reviewed their national pesticide disposal problems was perhaps the least successful segment of the course. Participants either had not had time to research the status of obsolete pesticide stocks in their countries, or had not been provided instructions in advance explaining the intent of the presentations. This should not be a problem for future deliveries of the course, provided there is sufficient time in advance of the workshop to notify participants and allow them time to research and prepare for the presentations.

Participants were very interested in the disposal of small quantities of pesticides and the handling of small containers (e.g., in amounts used by farmers) - subjects not covered by the course. The second delivery in Central America will be modified to include time to address this concern, and future deliveries in other regions will incorporate opportunities to discuss regionally-significant disposal issues.

Indonesia — December 1997

In December 1997, an EPA team delivered the course in Indonesia. The course was held at Wisma Taman Indah, a facility in Cisarua, which is about 60 km south of Jakarta. Twenty-seven attendees from Indonesia participated, including civil servants from the Indonesian Pesticide Commission Secretariat, plant protection specialists/ extension agents from thirteen of the Indonesian provinces, and representatives from FAO.

What Worked. The translation services were outstanding. All three of the FAO documents and all of the overheads for the lectures had been translated into Bahasa Indonesia prior to the course and copies were provided to the participants.

Two FAO videos — on disposal operations in Niger and Yemen — were included in the curriculum for the first time. The videos were a great addition to the course because they provided visual and "real life" examples of the procedures, equipment, and other information discussed in the lectures.

The case study provided an enjoyable and successful mechanism for attendees to apply the information learned in the course. The scenario involved three storage areas with potentially obsolete pesticides in the hypothetical country of Mayapada. The participants were grouped into three teams competing to present the best solution for the country's pesticide disposal problems to the "Mayapada Minister of Agriculture." The attendees worked diligently and gave three good, polished presentations after lunch on the third day, finishing the case study much earlier than participants in the previous two deliveries of the course. The three groups also presented an overview of the causes of the accumulation of stocks of obsolete pesticides in Indonesia.

In order to provide a qualitative evaluation of the effectiveness of the course, participants were asked to answer the same two questions before the course and then after the course. The questions asked about the person's understanding of (1) the steps and procedures involved with a successful pesticide disposal operation and (2) the advantages and disadvantages of different disposal options. Participants were given the options of "no," "little," "good," or "full" understanding. Prior to the course, the vast majority of participants claimed "little understanding" for both questions. In the post-course questionnaires, the vast majority of participants claimed a "good understanding" in response to both questions (*12*).

What Didn't Work. As in the Honduras workshop, participants were very interested in the disposal of small quantities of pesticides and the handling of small containers (e.g., in amounts used by farmers). This again reinforces the need to include time for discussion about issues important to the region.

Asking leading questions about specific topics that had just been introduced generated much less discussion than did recording ideas on a flip chart. The lack of success with leading questions can be attributed at least partly to the language barrier; often it was not clear that the lecturer was asking the participants a question, and translation was needed. Another consideration, though, is that a brainstorming session where many people are tossing out ideas is a much less threatening form of participation to most people.

EPA's hosts from the Indonesian Secretariat of Pesticide Commission made arrangements for the course attendees to visit a nearby pesticide warehouse on the fourth day of the course. However, the owners of the warehouse reconsidered their invitation shortly before the course began, and the visit had to be canceled. A visit to a new state-of-the-art hazardous waste disposal facility was quickly arranged for that day. While the tour of the disposal facility was useful for illustrating a possible disposal method for obsolete pesticide stocks, it did not provide the attendees the experience of taking an actual inventory, nor an illustration of potential difficulties in stabilizing and cleaning a warehouse of obsolete pesticides.

Changes Made

Since the May 1997 delivery, the EPA Office of Pesticide Programs has been refining lecture material, slides, and improving interactive methods to teach the course fundamentals.

The presentation on the proper use of personal protective equipment (PPE) is an example of how EPA is changing the course to make it more engaging for participants. During the dry-run of the course in April 1997, the PPE material was presented in a series of slides. For the first delivery in Honduras, an EPA team member dressed up in the PPE, complete with respirator and gloves. Some components of PPE were missing or used incorrectly, and participants helped identify and correct the problems. This modification has been incorporated into the course.

Based on the comments received and how the lectures flowed, EPA is modifying all of the lectures prior to the next delivery. For example, it has been strongly recommended that we include regional-specific details in the "Legal and Logistical Considerations" lecture.

Module as a Tool to Manage Global Risks

The "Pesticide Disposal in Developing Countries" course is a useful tool for to help manage global risks by working at a personal, regional and global level. EPA's intention is to provide decision-makers a technical and legal framework for regional action to eliminate obsolete pesticide stocks. By informing participants of the risks associated with obsolete pesticides, the course is a call to action. By training individuals to be facilitators and to take ownership of the course, it becomes a self-sustaining instrument for the dissemination of information to those involved in the regulation and disposal of pesticides, and a foundation for building a network of regional contacts.

The course works well as a component of larger, more comprehensive projects. For instance, the course in Honduras was part of a technical assistance agreement first established in 1991 to address "circle of poison" concerns of the US public and the desire of the Central American countries to export agricultural commodities to the US. At the request of the Central American countries, this project was expanded into a multi-step process to dispose of obsolete stocks. The following steps have been proposed or completed:

Step 1: Delivery of the "Pesticide Disposal in Developing Countries" course in Zamorano, Honduras in May, 1997.

Step 2: Second delivery of the course in Central America, tentatively planned for autumn 1998 at OIRSA headquarters in El Salvador. Two participants from the first delivery will join an EPA team in facilitating the course. Preliminary inventories have been completed for Belize (3 tons identified), Honduras (212 tons), and Nicaragua (1,665 tons). Participants in the second delivery have been asked to prepare preliminary inventories for their countries in advance of the course.

Next steps: A meeting is planned for late 1998 in Central America to develop a regional portfolio of proposals for pesticide disposal operations. These proposals will be based in part on the preliminary inventories of obsolete stocks produced for the Disposal Course. Then, with EPA assistance, our regional counterparts will sponsor a meeting in early 1999 with representatives from donor agencies, development banks, and private industry to review the portfolio of disposal proposals and identify financial assistance that would be available for implementing some of the projects. Working regionally or individually, countries will then be able implement their national disposal plans.

Conclusion

The course "Pesticide Disposal in Developing Countries" was developed by EPA as a potential tool to help manage the local, regional, and global risks associated with stockpiles of obsolete pesticides, especially in developing countries and countries with economies in transition. Through classroom lectures and practical exercises, participants are familiarized with the risks of inaction, given an overview of how to control risks during a disposal operation, and encouraged to champion practices that can prevent the accumulation of future obsolete pesticide stocks. This course was successfully delivered in Honduras and Indonesia in 1997; a second delivery in Central America is planned in conjunction with a larger pesticide disposal project.

Literature Cited

1. FAO. Informational Note for the Food and Agriculture Organization of the United Nations Third Consultation on Prevention and Disposal of Obsolete and Unwanted Pesticide Stocks in Africa and the Near East, Rome, Italy, March 2-3, 1998.

2. Stanislaw, S. "Progress and Developments on Unwanted Pesticides in Poland," included in the proceedings from the 4th Forum on HCH and Unwanted Pesticides, Poznan, Poland, 15-16 January, 1996.
3. Jones, M. Final Trip Report by Margaret Jones, US Environmental Protection Agency Region 5, Chicago, IL, " USAID Pest and Pesticide Management Project in Ukraine: Institutional Strengthening Component, Puscha Vodytsia and Kiev, Ukraine", 1997.
4. FAO. *Obsolete Pesticides: Problems, Prevention and Disposal*; Food and Agriculture Organization of the United Nations, Rome, Italy, 1998.
5. Jensen, J. Trip Report, "FAO Experts Meeting on Disposal of Obsolete Pesticide Stocks in Africa and the Middle East, Rome, Italy", 1998.
6. FAO. Provisional Technical Guidelines *Disposal of Bulk Quantities of Obsolete Pesticides in Developing Countries*; Food and Agriculture Organization of the United Nations, Rome, Italy, 1996.
7. FAO. *Pesticide Storage and Stock Control Manual*; Food and Agriculture Organization of the United Nations, Rome, Italy, 1996.
8. FAO. Provisional Guidelines on the *Prevention of Accumulation of Obsolete Pesticide Stocks*; Food and Agriculture Organization of the United Nations, Rome, Italy, 1995.
9. EPA. Informational Catalogue on International Training Modules, EPA-B-96-002, US Environmental Protection Agency, Washington, DC 20460, 1996.
10. Dwinell, S. Assessment of Pesticide Disposal Activities in Nepal, Final report on Second Mission, for the Asian Development Bank, Manila, Philippines, 1992.
11. Costello, K., and J. Jensen. Trip Report, "Course on Pesticide Disposal in Developing Countries, El Zamorano, Honduras", 1997.
12. Fitz, N. Trip Report, "Course on Pesticide Disposal in Developing Countries, Cisarua, Indonesia", 1997.

AUTHOR INDEX

Author Index

SUBJECT INDEX

Subject Index

H

N

R

T